T0398816

THE HUNDRED YEARS' TRIAL

The HUNDRED YEARS' TRIAL

Law, Evolution, and the Long Shadow of *Scopes v. Tennessee*

Alexander Gouzoules and Harold Gouzoules

JOHNS HOPKINS UNIVERSITY PRESS BALTIMORE

 Published 2025
Printed in the United States of America on acid-free paper
2 4 6 8 9 7 5 3 1

Johns Hopkins University Press
2715 North Charles Street
Baltimore, Maryland 21218
www.press.jhu.edu

Library of Congress Cataloging-in-Publication Data

Names: Gouzoules, Alexander, 1986– author. |
Gouzoules, Harold, 1949– author.
Title: The hundred years' trial : law, evolution, and the long shadow
of Scopes v. Tennessee / Alexander Gouzoules and Harold Gouzoules.
Description: Baltimore : Johns Hopkins University Press, 2025. |
Includes bibliographical references and index.
Identifiers: LCCN 2024052003 (print) | LCCN 2024052004 (ebook) |
ISBN 9781421452173 (hardcover) | ISBN 9781421452180 (ebook)
Subjects: LCSH: Scopes, John Thomas—Trials, litigation, etc. |
Evolution (Biology)—Study and teaching—Law and legislation—United States. |
Evolution (Biology)—Religious aspects—Christianity. | Religion and science—
United States—History. | Racism in textbooks—United
States—History—20th century.
Classification: LCC KF224.S3 G68 2025 (print) | LCC KF224.S3 (ebook) |
DDC 344.73/077—dc23/eng/20241105
LC record available at https://lccn.loc.gov/2024052003
LC ebook record available at https://lccn.loc.gov/2024052004

A catalog record for this book is available from the British Library.

Special discounts are available for bulk purchases of this book. For more information, please contact Special Sales at specialsales@jh.edu.

EU GPSR Authorized Representative
LOGOS EUROPE, 9 rue Nicolas Poussin
17000, La Rochelle, France
E-mail: Contact@logoseurope.eu

For Sophie

CONTENTS

PREFACE

THE CENTENNIAL OF THE 1925 SCOPES "monkey trial"—formally, *State of Tennessee v. John Thomas Scopes*—is a fitting moment to reflect on the enduring significance and profound impact of those distant proceedings on the cultural, legal, and educational landscapes in the United States. Scopes, a high school teacher in Dayton, Tennessee, was charged with violating a state law banning the teaching of evolution in public schools. The ensuing trial was far more than a legal battle over the teaching of evolution and the theory's implications for human origins. It was that, but *Scopes* has also come to symbolize a crucial juncture in the American narrative, where the intertwining threads of science, religion, law, and media converged to shape public discourse. It was a pivotal moment in US legal and educational history, and it retains interest in part because the results and legacy remain ambiguous.

At its core, the *Scopes* trial became emblematic of the cultural and scientific shifts occurring in the United States during the early part of the twentieth century. The widespread acceptance of evolution, if not the mechanisms underpinning it, within the scientific community was increasingly at odds with traditional biblical interpretations of the origins of life, especially in more religious and politically conservative sections of the country. These tensions went far beyond academic debate, penetrating deeply into public concerns about morality, democracy, education, and the role of religion in American life. Many looked to *Scopes* for answers to broader questions about academic freedom, the separation of church and state, the relationship between courts and legislatures, and the place of scientific theories in the curriculum of public schools.

What makes the anniversary of the *Scopes* trial especially noteworthy is its continued relevance in the modern world. The issues at the heart of the trial—conflicts between science and religion, controversies about how to educate the public on scientific developments, and unease over the power of courts to override democratic legislation—are as pertinent now as they were a century ago. The trial and its legacy are sobering reminders of the challenge faced by modern educators tasked with the public's education about areas of scientific consensus, including climate change, vaccine efficacy, and—yes, still—evolution. These contemporary discussions echo the same questions about the very nature of science, the extent of democratic control over educational content, and the influence of religious beliefs on public policy.

Academically, the *Scopes* trial has been a subject of continuous interest, analyzed as an inflection point in the evolution of educational curricula, American political ideologies, and media culture. Sitting at an intersection of multiple disciplines, the trial provides a rich case study for scholars. It offers insights into how legal battles can both arise from and become focal points for broader societal conflicts—and how these conflicts can then shape the trajectory of legal and political reforms.

The legacy and subsequent history of *Scopes* is no less interesting. The trial's immediate aftermath saw a lull in legal disputes over evolution, largely because evolution was subsequently deemphasized in public education. But many analogous cases concerning evolution and the school system came later, with more than one subsequent case billed by the media as "*Scopes* II." These have twice reached and been decided by the United States Supreme Court and have been litigated in district courts across the country (often, but by no means exclusively, in the South). Indeed, the last full decade *without* significant litigation over the place of evolution in the public school system was the 1950s. *Scopes* and its legacy thus also provide fascinating insight into the inability of seemingly definitive legal decisions to resolve with finality continuing social, cultural, and political controversies.

Naturally, then, many of the themes and topics we explore in *The Hundred Years' Trial* have received considerable attention in previous works: the *Scopes* trial itself; the history of ideas concerning evolution, including controversies about human origins and tensions with religious ideas; internal debates within the science of evolution, and how those debates have been received and understood by the public; the place of evolution controversies in the development of US constitutional law; the intersection of science, politics, and education; and the biographical details of key players populating the *Scopes* story. All of these are explored here, but we of course acknowledge that our exploration follows those of many others.

In particular, historian Edward J. Larson's masterful *Summer for the Gods*[1] has been the gold standard for comprehensive and integrated coverage of *Scopes*—but its temporal radius is narrower than what we imagined for this new effort. Larson's account of the events leading up to the trial examines American societal trends in the early twentieth century. Here, we reach for a more distant starting point, beginning our story with Darwin, the emergence of modern evolutionary theory, and the immediate controversies that followed and intensified as evolutionary ideas crossed the Atlantic. The status and impact of the *Scopes* trial, we suggest, were spawned by forces sparked by Charles Darwin in 1859, with the publication of *The Origin of Species*. Our story begins with that transformative event.

We also devote a large portion of our analysis to post-*Scopes* developments, such as the scientific synthesis of evolutionary theory, emerging post-synthesis controversies within the evolutionary field (and the extent to which those controversies mirror those that existed in the 1920s), and significant evolution-in-schools cases like *Kitzmiller* (a federal challenge to a Dover, Pennsylvania, school district policy that required intelligent design content in the curriculum), which were litigated after *Summer for the Gods* was published.

We believe there is ample room for a new assessment of the legal legacy of the *Scopes* case, as the Supreme Court's interpretation of the Constitution's religion clauses has changed dramatically and fundamentally since 1997. Larson wrote—with good reason at the time—that the Supreme Court's jurisprudence of the 1960s "finally provided solid authority for effectively challenging antievolution statutes under the federal Constitution [and the] Scopes legend did the rest." Today, much of that legal authority is now gone. The recent overruling of the Supreme Court's venerable *Lemon* test has removed a foundational precedent relied on in the majority of the many post-*Scopes* evolution decisions, and a Supreme Court now dominated by appointees of George W. Bush and Donald Trump has decried foundational religion law cases as "abstract and ahistorical." This reversal has been followed by initiatives to return religion to the public school curriculum by way of mandatory Ten Commandments displays, biblical instruction, and measures to cast doubt on evolutionary science. These new developments are covered here.

In addition to Larson, biologist Randy Moore also stands out among scholars of the *Scopes* trial. His research in the field of biology education, particularly the teaching of evolution, has examined *Scopes* extensively.[2] *The Hundred Years' Trial* extends and complements Moore's efforts with its integration of an extended history of evolutionary biology with a thorough analysis of the development of constitutional legal issues implicated by the *Scopes* trial, presented in the context of evolving jurisprudential philosophies and methodologies.

Given *Scopes*'s importance to the nation's history, it is unsurprising that we are not alone in recognizing the significance of its centennial. Two new books on the trial have been published as ours went to press.

The Trial of the Century,[3] by Gregg Jarrett with Don Yaeger, is a colorful and straightforward retelling of the trial. Jarrett is a legal and political analyst for Fox News, and co-author Don Yaeger is a writer and associate editor at *Sports Illustrated. The Trial of the Century* does not break new ground in its retelling of *Scopes,* but the book does stand out as a paean to Clarence Darrow, whom Jarrett admired since childhood. The conclusion that despite Scopes's conviction, Darrow "ultimately prevailed in his efforts to preserve intellectual freedom and the advancement of science," and that the story reveals the need to unite in the defense of the free exchange of ideas, reflects the traditional, midcentury narrative about the case. However, our account emphasizes the complications, subtlety, and nuance of the trial, its background, and its 100-year aftermath.

Finally, literary critic and essayist Brenda Wineapple's *Keeping the Faith: God, Democracy, and the Trial That Riveted a Nation*[4] masterfully captures the dramatic essence and cultural significance of the *Scopes* trial, presenting it as a profound clash of ideologies that still resonates today. Her work emphasizes cultural conflicts that the debate over evolution symbolized rather than the science behind evolution, and she views the trial as a stand-in for broader societal tensions and cultural wars over America's identity. We believe *The Hundred Years' Trial* complements Wineapple's focus on cultural exploration by delving deeper into the scientific aspects of evolution and their interactions with a legal system not always well suited to evaluate science. We believe our discussion illuminates the way scientific debates have continued to evolve (and sometimes ignite) in American society over the past century. We argue that although cultural conflicts certainly framed the *Scopes* trial and its immediate aftermath, the science of evolution itself has played a crucial and enduring role in shaping public discourse, scientific understanding, and educational policy.

Our goal with *The Hundred Years' Trial* is to provide readers with a new analysis that blends fresh and topical legal, biological, and psychological perspectives on *Scopes*. We have sought to achieve this by melding our different areas of expertise and academic training. Alex is a constitutional scholar who has litigated First Amendment cases implicating the separation of church and state (including those arising from controversies related to religion in public schools); he teaches law at the University of Missouri. Harold is a biologist specializing in the evolution of animal and human communication systems who has taught evolutionary biology and its history for forty years at Emory University. As father and son coauthors, we have a tactical advantage in speaking with one voice despite these different academic backgrounds, thanks to our lifetime of conversations (and, probably, a shared love of Atlanta Braves baseball).

Having described our goals for the project, it is also worth briefly noting what we are not attempting. Given the vast array of disciplines that have engaged with *Scopes* and its legacy—law, history, education, biology, sociology, theology, media studies, even theater studies—no single effort can fully engage with every dimension while maintaining coherence. The history and theology of religious movements in the United States is largely outside the scope of this work. We explore and analyze the public reaction (including by those of various religious perspectives) to developments in evolutionary theory, and we attempt to explain the various aspects of evolutionary biology that triggered the vehement opposition of William Jennings

Bryan in particular. But these pages do not exhaustively catalogue the specific claims and contentions of the various anti-evolution movements that proliferated later. To the extent that the labels for these various ideas (e.g., "scientific creationism" and "intelligent design") reflect authentically different positions, rather than a strategic repackaging of the same core objection, others are better equipped to (and have) analyzed them.[5]

The largest challenge we faced, given the enormous literature available on the different dimensions of the *Scopes* saga, was condensing the story into a trim volume. We drew courage from historian Janet Browne, who devoted over a thousand pages to her remarkably rich two-volume biography of Charles Darwin, yet was able to distill the core content into a subsequent, much slimmer 174-page account of the great evolutionist and his transformational book, *The Origin of Species*. Browne's talent and depth of scholarship are inspirational.

Far from a historical relic, the *Scopes* trial remains relevant in the continuing narrative of the complex relationship among science, law, religion, education, and democracy. Its centennial is an important opportunity to reflect on the nature of public discourse from 1859 to the present. This is an appropriate time to engage in reflective discourse, reassess our perspectives on the issues raised in 1925, and appreciate the trial's contribution to the ongoing dialogue about the role of education in a democratic society. We hope *The Hundred Years' Trial* contributes to this conversation.

THE HUNDRED YEARS' TRIAL

Introduction

A VANISHINGLY SMALL NUMBER of court cases are still remembered, at least outside the realm of legal scholarship and education, a hundred years after their conclusion. A few major Supreme Court decisions make the cut—the foundational *Marbury v. Madison* and the notorious *Plessy v. Ferguson* are usually taught and sometimes remembered as significant parts of American history.[1] Some infamous murder cases make the cut as well. The trials of Bruno Hauptmann (the killer of the Lindbergh baby), Leon Czolgosz (the assassin of President McKinley), or Leopold and Loeb (brutal teenage murderers from elite families) retain some historical interest independent of the underlying actions of the defendants. These are all exceptions, not the norm.

The 1925 trial of Tennessee high school teacher John Scopes, charged under a new statute criminalizing the teaching of evolution in a public school, is a curious member of this exclusive list. A misdemeanor trial in a small town courtroom, resulting in a hundred-dollar fine later set aside on a technicality, would seem destined for historical irrelevance. Nonetheless, the case remains "a staple in every US history survey class."[2] It is regularly examined as background in modern court decisions and legislative debate, especially in disputes involving religion and education.[3] It has also served as a tailor-made analogy for contemporary clashes over the science of climate change[4] and the medical effectiveness of vaccines.[5] The media, too, continues to invoke the case, having billed several subsequent trials—and there have been *many* involving the proper place of evolution in school curricula—as "*Scopes* II."[6]

The *Scopes* trial has maintained a hold on the minds of scholars as well. Over the years, many have added complexity to the traditional explanatory narrative of the trial, which framed these events as a conflict between fundamentalist religion and modern science.[7] More recently, the *Scopes* trial has been reevaluated through the lens of race and gender, examined as media spectacle and "circus" politics, reframed as a conflict over textbooks, re-positioned as an inflection point in an ideological transition from an older majoritarian progressivism to a newer liberalism focused more intently on individual liberty, and contextualized as a part of the wider progressive struggle against social Darwinism and legal formalism.[8]

Given this durable cultural and academic legacy, the facts of the strange case are generally familiar—but a brief overview may nevertheless serve as a starting point for discussion, assessment, and evaluation of its unique legacy.[9]

The early twentieth century saw significant cultural change and scientific development, which in turn produced their own discontents in the United States. These upheavals led to a cultural and theological clash between religious fundamentalism and modernism; this backlash was especially intense regarding accounts of evolution, which offered a materialistic (rather than metaphysical) explanation for the differences between humans and other animals. Although the evolutionary field was in a state of internal disagreement over the mechanisms through which evolution takes place, with Darwinian explanations in temporary decline, the occurrence of evolution had gained overwhelming acceptance among scientists and educators. The result, at a time when compulsory public education was relatively new to some areas of the country, was widening exposure to a theory that traditionalists saw as standing in opposition to core biblical teachings about the relationship between man and God.

In response to fundamentalist campaigns, several states, mostly in the South, adopted prohibitions against the public instruction of evolutionary theory as applied to humans. A young and ambitious American Civil Liberties Union (ACLU) organized a challenge to Tennessee's novel anti-evolution law, known as the Butler Act. The ACLU recruited Dayton, Tennessee, high school teacher John Scopes to volunteer as a defendant. This plan to test the constitutionality of a new anti-evolution law drew the interest of the law's most significant champion, William Jennings Bryan, a three-time Democratic nominee for president and former secretary of state. Bryan signed on to assist the prosecution and would overshadow the government attorneys working alongside him. Not to be outdone by a former ally turned rival, star defense attorney Clarence Darrow, a prominent champion of liberal causes, joined the defense. Thus, the stage was set for a witness-stand showdown between the two that would pass into legend.

The prosecution viewed the legal issues presented by the case as simple ones: a state law stood on the books, and Scopes had broken it. Whether or not that law was wise or accurately reflected the modern scientific consensus were questions to be resolved by the democratic process, not by judges or juries. Here, the prosecution's vision of the case echoed arguments that had long been advanced by turn-of-the-century progressives like Bryan, a key figure in a dynamic movement that had championed state legislative reforms, only to have them stymied through judicial review by a largely conservative legal system.

The defense, on the other hand, planned to introduce the testimony of a collection of scientific experts, brought from around the country to explain—to the jury and to the world—the fundamental importance of evolutionary theory and the broad consensus as to its validity. In 1925, legal appeals to scientific consensus, data, and analysis were a relatively recent and uncommon phenomenon, and the formal relevance of evolutionary theory's scientific validity to the constitutionality of Tennessee's law was not self-evident. The state's team convinced the sympathetic Judge Raulston to exclude most expert testimony from the jury. Much of the defense's carefully collected scientific testimony was placed in the trial transcript for publication and review on appeal, but it was not heard by the Dayton jury.

Despite a massive presence by the national media and the attending publicity, Raulston's limiting ruling seemed to condemn the trial to be a dud—until Darrow shrewdly called Bryan as an expert witness on the Bible. It was an unorthodox legal maneuver, inexplicably allowed by a judge who had quarreled with the agnostic Darrow for days over the propriety of opening court sessions with prayer. (The defense team had lost that particular battle, but they had at least convinced Raulston to remove from the courthouse a sign urging onlookers to "read your bible.") The examination was also a continuation of a long-running debate between the two attorneys over differences between Darrow's agnosticism and Bryan's fundamentalism, which had led them to opposite positions on divisive issues like prohibition and evolution.

The ensuing witness stand debate between Darrow and Bryan over biblical contradictions—a topic far removed from the legal questions presented in the case—was breathlessly covered in national newspapers and radio broadcasts, highlighting the trial's hybrid nature as a criminal proceeding and a mass entertainment spectacle.[10] After capturing the attention of the country, the theatric argument between two of the era's most famous men was finally shut down. Scopes was convicted and ordered to pay a token fine. Bryan died in Dayton soon afterward, with the strange case generally remembered as the nadir of his unique public career. The ACLU, in turn, failed in its goal of turning the case into a significant precedent for academic or religious freedom; instead, the Tennessee Supreme Court overturned the conviction based on a procedural defect, thus ending the appeal.[11] No one had gotten exactly what they wanted.

What accounts for the long shadow of these peculiar events? Many partial explanations are apparent but fail to wholly convince. For one, *Scopes* dominated American news coverage in the summer of 1925. The case even received significant (often condescending) attention in overseas publications. For example, in the revealingly titled article "The Case Against America," the literary supplement to the *Times of London* scoffed that "no European State could conduct a Scopes trial, because European peasants are not at one and the same time remote from the stream of ideas and surrounded by their material consequences."[12] But this alone cannot be enough. Similar media spectacles followed other so-called "trials of the century." Journalist H. L. Mencken, whose biting, cynical, and distinctive writing helped memorialize the *Scopes* trial, also sensationalized Hauptmann's case, for one, provocatively calling it "the greatest story since the Resurrection."[13]

Another potential explanation is that *Scopes* stands apart because it featured a dramatic confrontation between two prominent public figures. But Darrow litigated many other famous cases, including his controversial defense of the heinous murderers Leopold and Loeb, just the year before *Scopes*. Bryan, on the other hand, is not remembered for his skill as an attorney. (Indeed, he saw a need in his memoirs to defend against the charge that he had been an unsuccessful lawyer.)[14] Moreover, the two men had engaged in similar arguments about religion and science before, albeit on newspaper pages rather than in a courtroom.

The influence of Hollywood is another possible explanation for the longevity of *Scopes*. The trial's legacy was surely bolstered by the successful 1955 play, turned into a 1960 Hollywood movie, *Inherit the Wind*.[15] There, the *Scopes* case served as an analogy for the demagoguery of Joseph McCarthy during the second Red Scare. But even this explanation is not entirely satisfactory. Now, the themes of anti-communist overreach that permeate *Inherit the Wind* seem more culturally distant than does the case's original historical context. Today's America again features cultural conflict between the cosmopolitan national media and overlooked rural populations, charged debates over scientific conclusions (climate change and vaccine science, among others), recurring anger over content taught or excluded from public school curricula, and renewed progressive criticism directed toward judicial interference with the democratic process. Surely, it is the 1925 trial and its attending issues, rather than its 1955 re-imagining as a stand-in for McCarthyism, that continues to resonate.[16]

These contemporary parallels themselves account for the continued legacy of *Scopes* in American cultural and intellectual life—and justify a reassessment of the case at its centennial anniversary. In these chapters, we seek to contribute to explanations of the trial's long shadow by re-situating it within parallel stories of change in evolutionary biology and US law. Although many modern analyses have added important complexity to traditional narratives about *Scopes*, here we re-center the legacy of the trial on evolution itself, an idea which, from its development by Charles Darwin in 1859, down through the years to the present, has intrigued, provoked, tantalized, bemused—and yes, frightened—the human mind.

And it is an idea that has found its way with surprising frequency into US courtrooms. Every decade from the 1960s until today has featured at least one significant litigated case challenging the proper place of evolutionary theory in public school classrooms. To date, two of these cases have been decided by the US Supreme Court—normally an event that, at least in theory, adds settled finality to a previously uncertain area of law. Not so with evolution. As we argue herein, the legal status of evolution in US schools is less certain today than it has been at almost any time since *Scopes* itself.

If the Butler Act had barred the teaching of (for example) the history of the socialist movement in the United States, rather than the teaching of human evolution, any prosecution may well have been largely forgotten. This was, in fact, the fate of a little-remembered contemporary case, *Meyer v. Nebraska*, involving a World War I-era law banning German-language instruction in public schools.[17] But whereas *Scopes* warrants discussion in survey American history courses, *Meyer* is generally relegated to law school electives.

Evolution is a remarkable idea, at once foundationally important, challenging to various belief systems, and built upon a complex body of evidence that is difficult to decipher and easy to mischaracterize (either unintentionally, through misinformation, or intentionally, through disinformation). As such, it has often been beyond the legal system's capacity to effectively grapple with. Some cases that have engaged with evolution, including *Scopes* as well as analogous successor cases like *McLean v. Arkansas*[18] and *Kitzmiller v. Dover*,[19] featured extensive expert testimony by some of the world's leading evolutionary biologists. Many others have been decided with minimal scientific testimony. Our review of this history finds little evidence of scientific presentations about evolution affecting case outcomes.

This conclusion raises important questions of its own: What role *should* the robust scientific consensus that evolution occurred (or, for that matter, the many recurring internal debates regarding the mechanisms and processes through which it occurred) play when questions about evolution are raised in a courtroom? And to the extent that science has a place in legal debates over evolution, why has it generally failed to make much of an impact?

Our analysis tracks legal challenges to evolution alongside the extended, nuanced, and recurring controversies that have accompanied developments within the evolutionary field. It positions *Scopes* as the first, but by no means final, attempt to challenge the evolutionary consensus from within a courtroom. Once that attempt was made at *Scopes*, the effort never truly ended.

Our analysis has noteworthy implications both within and beyond the context of evolutionary biology. *Scopes* and its recurring successor cases point to challenges faced by evolutionary biologists in explaining their field's conclusions to outside audiences. At the same time, these cases demonstrate the difficulties the legal system faces when it addresses scientific conclusions that are based on complex and multifaceted bodies of evidence. In this, seemingly irresolvable legal debates over evolution may presage important cases involving climate change, another field where experts have been hard-pressed to convincingly explain complex bodies of evidence to a skeptical general public as well as to legal decision makers.[20]

First, our account will lay the groundwork for *Scopes* and its long aftermath. To explain why the idea of evolution was so polarizing—both in 1925 and in later years—it begins earlier, with Charles Darwin's pioneering and controversial ideas. A famous passage from Darwin's *The Origin of Species* says much on its own:

> There is grandeur in this view of life, with its several powers, having been originally breathed into a few forms or into one; and that, whilst this planet has gone cycling on according to the fixed law of gravity, from so simple a beginning endless forms most beautiful and most wonderful have been, and are being, evolved.

Grandeur, indeed, but also mystery, intrigue, and heated controversy. Many scientific developments (geological evidence as to the age of the Earth, for one), stand in tension with the most literal reading of biblical passages. But only evolution has given rise to a hundred years of contentious litigation over its place in public science curricula. We thus begin our story with Darwin, for the path to Dayton begins in Down, the English village that was Darwin's home.

We next trace the initial reactions, objections, and controversies that were sparked by Darwin's work. We chart the reaction to evolution as well as the way evolutionary theorists responded—sometimes by making precise rebuttals, but other times by amending Darwin's theory in counterproductive directions. We analyze these discussions of evolution as they began in Britain and spread to the United States. In these early chapters, we also examine two distinguishable but related intellectual developments with dark legacies: social Darwinism and eugenics. Although these ideas were bit players in the *Scopes* trial itself, some discussion is nonetheless relevant, both because concerns about them motivated some at the time to oppose evolution, and because, in the subsequent years, exaggerated links between these ideas and evolutionary biology have been deployed by Bryan's many successors.

Our discussion thus advances to the lives of Bryan and Darrow themselves, using their fascinating and intertwined careers to discuss the state of US law and society in the era leading to *Scopes*. We discuss how several trends during this era paved the roads they would walk while laying the foundation for *Scopes*, including heated questions about the appropriate level of deference that judges should show to legislatures, innovative approaches to legal argument that emphasized the presentation of empirical and scientific data, and controversies about which individual liberties are safeguarded by the Constitution against infringement by state legislatures. We trace the rise of the anti-evolution movement and the enactment of the Butler Act in the context of this complex era of US history.

We then turn to the *Scopes* trial itself. Although these legendary events have been covered before, our emphasis is twofold. We focus on the legal arguments presented by the *Scopes* defenders, showing how they were at once innovative for the time and also flawed. We also discuss the scientific evidence marshaled by the defense, which provides a fascinating insight into the state of the evolutionary field on the cusp of the transformation known today as the Modern Synthesis. At the time of *Scopes*, Darwinian natural selection had not yet won out over competing evolutionary theories such as mutationism, and the result was a potentially confusing corpus of expert testimony.

After covering Bryan's death and the *Scopes* appeal, we then follow the *Scopes* legacy through the 1930s, 1940s, and 1950s—the only three decades since the *Scopes* trial during which no significant legal case involving evolution was litigated. But these years were critically important: we discuss how the remarkable intellectual movement known as the Modern Synthesis brought the evolutionary field back into stability by answering most of the pending questions about evolutionary mechanisms that had lingered in the 1920s.

The legacy of *Scopes* appeared set by the midcentury. Around the time that *Inherit the Wind* cemented the impression that *Scopes* had been, at best, a pyrrhic victory for Bryan's movement, a Cold War push for scientific competitiveness injected the newly confident field of evolutionary biology back into public education. In this context, the Supreme Court finally heard a challenge to the still-extant laws criminalizing evolution instruction in a handful of states. In *Epperson v. Arkansas*,[21] the Court held that such laws violated constitutional protections for the separation of church and state. Shortly thereafter, the Court signaled that federal courts would more actively enforce that separation by adopting the influential *Lemon* test, which governed cases involving public entanglement with religion.

Our narrative documents twin developments in the following decade. First, a series of attempts by state legislatures to refine and enact new anti-evolution laws to attempt to pass the *Lemon* test and avoid falling afoul of *Epperson*, each followed by a new round of litigation that echoed the controversies and dynamics of *Scopes*. Second, we document a series of internal controversies within the field of evolutionary biology in the wake of the synthesis. Much like the scientific controversies that were ongoing in 1925, these involved debates over evolutionary processes and mechanisms—not pushback or uncertainty about the fact that evolution occurred. Nevertheless, this contributed to the many efforts to challenge evolution instruction in court by exacerbating public misunderstandings about the state of evolutionary theory.

We ultimately argue that the *Scopes* legacy remains in flux, with consensus peaking in the midcentury. As we demonstrate, continuing court cases involving evolution have become more contentious—moving from the relatively amicable, circus atmosphere that dominated *Scopes* into a hostility that permeated the more modern cases, some of which sparked threats to the safety of the participants. We also argue that the rise of more conservative philosophies of jurisprudence, coupled with a dramatic change in the makeup of the Supreme Court, has put the legal foundation of the post-*Scopes* consensus on shaky ground. The *Lemon* test was recently overturned, and the Court has suggested that other religion-law precedents, which may include *Epperson*, may follow. At the same time, in cases involving, for example, healthcare and environmental science, the Court has demonstrated disinterest in evidence of scientific consensus. These developments and others clearly signal that the debate that consumed the nation one hundred years ago remains as relevant as ever.

But even the arguments at *Scopes* were themselves a continuation of a conversation begun by Darwin. We thus begin with him.

CHAPTER ONE

An Inordinate Fondness for Beetles

> May I, as one of the counsel for the defense, ask Your Honor to allow me to send you the *Origin of Species*?
>
> —Arthur Garfield Hays, Trial of John Scopes, Day 8

CHARLES DARWIN'S NAME was inextricably linked to evolution in 1925, and it remains so today.[1] Ask a college class to list prominent individuals in various academic disciplines—psychology, biology, philosophy, or physics—and many names will be thrown about. Do the same for evolution, and "Darwin" is the overwhelming response. Colloquially, "Darwinism" and "evolution" eventually became synonymous. And in 1925, Darwin's name was on the lips of the participants at *Scopes*.

In some ways, this seems perplexing. For evolution, Darwin's work was neither the first word nor the last.[2] The overall hypothesis that species arise through modifications of previous, ancestral forms did not originate with him. That idea and ones that brush shoulders with it date back to antiquity.[3] And importantly, the field we now know as evolutionary biology, although still focused on core issues established by Darwin, has expanded in ways Darwin could not have known.

Nevertheless, Darwin remains the pivotal figure in the evolutionary pantheon.[4] The intense controversies that burned in 1925 and continued long after were initially sparked by

Darwin's approach and position: His style made his dramatic scientific contributions broadly accessible, and therefore more of a lightning rod than they might have been had their initial presentation instead been esoteric. Additionally, his privileged class gave him access to an influential social network that amplified and publicized his discoveries. This unique combination transformed what might otherwise have been a more typical field of scientific study into a cultural, moral, and legal battlefield on both sides of the Atlantic.[5]

Critically, however, while Darwin's unique contributions inextricably link his person to the general idea of evolution, at the time of the *Scopes* trial Darwin's *specific* explanations had lost influence within the scientific community. This confounding state of affairs—a field tightly linked to a single man whose specific account had (for the time being) lost favor—contributed to a sense in the 1920s that evolution was both vulnerable to attack and required vociferous defense. These dynamics produced the *Scopes* trial, and they were rooted in Darwin's life. We thus begin with the story of how they arose.

ORIGINS

Charles Darwin was born into a well-to-do English family in 1809. His father, Robert Waring Darwin, and grandfather, Erasmus Darwin, were both doctors. Darwin later recalled that as a small boy, he was prone to inventing stories "for the sake of causing excitement."[6] For example, he once told a young friend that he was able to produce variously colored flowers by watering them with colored fluids. Charles was never punished for such fanciful fibs; his father was kind and gentle with his children and instead chose to shape Charles's behavior by refraining from overreaction.

In classical subjects such as Greek, Latin, and mathematics, the young Charles was an uninspired student. His passion was science, especially chemistry, biology, and geology. Here, he applied himself. He performed chemistry experiments with his brother in a homemade lab and was an avid reader, perpetually captivated by the natural world.

As a young man, Darwin had rare exposure to early ideas about evolution—most directly through the writings of his grandfather Erasmus, who had achieved considerable fame in his own right as a medical authority and early evolutionary thinker. Later, a brief stint as a medical student at Edinburgh University brought Darwin under the wing of Dr. Robert Grant, who had long been interested in Erasmus Darwin's ideas and was delighted to mentor his grandson. Grant then exposed Charles to the revolutionary theories of the French naturalist Jean Baptiste Lamarck.

Lamarck had challenged conventional ideas about science across fields such as geology and biology.[7] The traditional Western view, with roots tracing back to Plato, was that species were divinely created and immutable (unchanging). Both Lamarck and Erasmus Darwin challenged that view, with Lamarck forcefully arguing that aspects of animal behavior and other notable traits develop as responses to challenges presented by changing environments. Lamarck's model was that traits acquired by parents could be passed on to their offspring. Grant encouraged Charles to engage with these ideas.

However, although Darwin was expected to follow family tradition and become a doctor, he proved too squeamish for medicine. He left medical school for Cambridge University to become a clergyman, a career that would allow him to pursue his interests in natural history. At that point, he had no reason to—and did not—doubt the predominant creationist view of the origins of life. But early, direct exposure to the concept of biological evolution from Grant and his grandfather likely influenced his later ideas.[8]

By the time he arrived at Cambridge, Darwin had developed what would be a lifelong passion for natural history collecting, with a special fondness for collecting beetles. At university, he searched for these insects with friends during class breaks. On one occasion, he even placed one in his mouth so he could secure others with his hands. The trapped beetle released a noxious defensive spray, causing Darwin to lose all his specimens and undoubtedly amusing any witnesses. But even here, Darwin was on to something. Beetles proved to be of great interest to later evolutionary biologists because there are so very many species of them. The twentieth-century biologist J. B. S. Haldane, when asked what his distinguished research career had taught him about the nature of God, is said to have answered that He must have "an inordinate fondness for beetles."[9]

On his return from Cambridge, Darwin received word from his other academic mentor, botanist John Henslow, about an opportunity to become the ship's naturalist on a surveying vessel, the *HMS Beagle*. Darwin was not the only applicant for the position. But it was his good fortune to have family standing on his side (the position demanded it; it was unpaid), as well as a strong recommendation from Henslow, who had been offered the position himself but declined for family reasons.[10]

Before the *Beagle* set sail, Henslow recommended that Darwin take along the recently published first volume of Charles Lyell's *Principles of Geology*.[11] He cautioned Darwin to read it for its facts but to be cautious about the "wild theories" he would encounter in Lyell's book. But those theories, wild or not, would be of great interest. In *Principles*, Lyell presented a superbly argued case for understanding Earth's geologic past through natural laws and processes that remained active in the present. He emphasized gradualism—that geologic change was, almost always, exceedingly slow and built up over vast periods of time, rather than through cataclysmic and sudden occurrences. Darwin astutely saw parallels in the biological world. Life forms, too, might change over long periods of time due to natural processes.

On board, Darwin's duties were to collect and document natural history specimens—animals, plants, rocks, fossils—and send samples (literally tons of them) to Henslow in England for study and classification. This suited Darwin well, given his life-long interests in collecting and natural history. He threw himself into the job. This was no vacation cruise, but the young Darwin was a strong man with energy and vigor. When he was not seasick, he worked.

He also demonstrated strength of character. The ship's captain, Robert FitzRoy, had been eager to bring an educated naturalist onboard. FitzRoy was only four years older than Darwin and, as per custom, the two shared meals. This forced Darwin to deal with the captain's prickly and volatile personality, but he did so adeptly. At Bahia, Brazil, FitzRoy defended and praised slavery—which Darwin abominated[12]—by telling of a visit to "a great slave-owner" who had called up his slaves and asked them whether they wished to be free. The captain reported that their unified response was "no." Darwin countered by simply asking FitzRoy whether the response of slaves in the presence of their enslaver was meaningful. Annoyed, the captain banished Darwin from his cabin, and the young scientist feared he would be forced to leave the ship. But a few hours later (and not for the last time) FitzRoy sent an officer with an apology. The voyage continued.

The *Beagle*'s primary mission was to accurately map its route, especially poorly charted stretches along the South American coast. This gave Darwin opportunities for overland explorations in Argentina, Chile, and Brazil, where the tropical rainforest showcased abundant and exotic plants and wildlife. In an area south of Buenos Aires, he discovered remains of previously unknown species of extinct giant sloths and armadillos. It was clear to him that these creatures possessed the same basic body structure as modern versions still inhabiting the area. And yet, there were also noteworthy differences between earlier and extant forms.

Lyell's *Principles* had concluded that there was no known mechanism for the gradual transformation of animal types occurring in response to changing environments. Instead, Lyell contended that environmental change eliminated animal types that were no longer able to survive, and—somehow—new forms that fit the current state of the environment then appeared. Darwin's mind, however, was not closed on the matter. To him, the similarities between extinct and extant forms were too striking to ignore.

A pivotal moment arrived when the *Beagle* visited four islands of the Galápagos Archipelago, each within a hundred miles of each other.[13] Darwin was perplexed at the rich variation he observed across specimens from islands in so confined an area. Were variants of the same form (different finches or tortoises, for example) members of the same species, or of different species? Deeper questions about the very concept of species—and whether taxonomic boundaries, traditionally assumed to be sharp, might instead be fuzzy—began to occupy Darwin's thoughts. Was it possible that varieties within a species were "incipient species"? That extant species had themselves once been varieties?

With modern techniques, sequences of genes can today be identified and used to build phylogenies—trees of relatedness. Beyond DNA and RNA, mitochondrial sequences (which are relatively stable) also reveal ancestry. But even such state-of-the-art evidence must be interpreted carefully because convergent evolution can result in traits, and particular gene sequences, that are shared independent of ancestry. Thus, complementary information concerning geographical distributions and fossil record evidence is used to assist in the interpretation of gene sequences.[14]

Even with these cutting-edge techniques, establishing boundaries to determine species membership remains a complex and challenging task. The nebulous nature of the lines between species was one of Darwin's great insights, central to his view of evolution and his recognition of continuity and gradual change. Within a year of the *Beagle*'s return home in 1838, Darwin began to reject traditional notions about the stability of species.

By that time, his meticulous collections and detailed letters home helped cement his status and reputation in the scientific community. In this, he was also aided by his social position and notable ancestor, his grandfather Erasmus. Through his advantageous social circles, Darwin soon met Charles Lyell, author of the geology text he'd found intriguing and helpful, and the two became fast friends and confidants.

Nevertheless, Darwin did not publish his seminal work, *On the Origin of Species*, until 1859. Scholars have debated why it took some twenty-one years to publish his transformational ideas.[15] Once he put pen to paper, the final manuscript took only thirteen months and ten days to complete. But Darwin first devoted two decades to marshaling evidence and corresponding with colleagues and knowledgeable people across the world. He had planned an overwhelmingly convincing tome, but intervening events shaped its release and his approach.

COMPETITION AND ADAPTATION

The intervening period gave Darwin a roadmap of how *not* to proceed. Late in 1844, an anonymous and controversial book on life's origins, *Vestiges of the Natural History of Creation*,[16] received worldwide attention and interest—even Queen Victoria had it read to her. The author put forth the materialist (and potentially heretical) idea that all knowable existence—the solar system, Earth, and the life forms inhabiting it—arose from earlier forms through natural laws and processes without input from a creator. It was an all-encompassing argument for "transmutation" and was a bestseller, going through twelve editions. The last was published in 1884, altered from the eleventh only in that the identity of the author, the prominent publisher and journalist Robert Chambers, was finally revealed after his death.

Vestiges of the Natural History of Creation was the first widely available book to present to the general public the idea that extant species originated from earlier forms. Chambers had anticipated vicious and scornful criticism and thus proceeded anonymously. His book used a variety of evidence, including fossils and embryological comparisons across species, to argue (in a cleverly reverential tone) that living things progress from the simple to more complex over time. The progression Chambers proposed was linear (not branching, as Darwin's vision of evolution was to be), and Chambers proposed no mechanism to account for the phenomenon. Darwin read *Vestiges* carefully, soon after its publication. He was scathing in his criticism of its evidence, which he considered inaccurate and simplistic. He also kept careful track of the harsh reviews it received.

Other naturalists were reading as well. For instance, Alfred Russel Wallace, born when Darwin was fourteen and on the precipice of his brief foray into medical study, came from a humble and impoverished English family. He would never be positioned to assume unpaid duties as a ship's naturalist; as a young man, he earned a living as a land surveyor and teacher. But like Darwin, Wallace's real passion was natural history. He was intrigued by *Vestiges* and by Lyell's *Principles of Geology*, as well as the works of Thomas Malthus, which influenced his thinking about competition in nature. Wallace was eventually able to embark on natural history collection expeditions in the Amazon, and later the Malay Archipelago. In contrast to Darwin's voyage on the *Beagle*, he set off with a specific interest in finding evidence compatible with his budding idea of evolutionary change.

Wallace reached similar conclusions as Darwin at around the same time. That said, modern scholars have analyzed details of Wallace's conception of natural selection, finding that it holds up less well than Darwin's vision.[17] Nevertheless, Darwin panicked when he received a package from Wallace (who had read his published account of the voyage of the *Beagle*) before Darwin had published his own theory. The package included a copy of an article describing Wallace's ideas about evolutionary change and natural selection.

Concerned by this unexpected competition, Darwin went to Charles Lyell and another prominent confident, Joseph Hooker, who together came to his rescue. They recommended that Wallace's paper be introduced to the scientific community alongside early drafts of Darwin's work. The venue would be the July 1858 meeting of the Linnaean Society of London, and both papers would be published together in the Society's journal. Remarkably, this was arranged without consultation with Wallace, who was still in Malaysia. But in the end, he had no objections. Thus, Darwin and Wallace became formal co-discoverers of natural selection.

The last-minute additions to the Linnaean Society session confused those in attendance but generated no comment when read. Thomas Bell, the Society's president, in reviewing the group's activities in his annual address in May 1859, declared that the past year had not "been marked by any of those striking discoveries which at once revolutionise, so to speak, the department of science on which they bear."[18] Few statements could be further from the truth.

REVOLUTION

Darwin finally published his great book in 1859, with the full title of *On the Origin of Species by Means of Natural Selection, or the*

Preservation of Favoured Races in the Struggle for Life. By this point, the idea that species change over time was by no means novel or revolutionary. But the centerpiece of *Origin* was Darwin's recognition of natural selection and its role in producing change, and new species, over generations of reproducing organisms. By the time of the *Scopes* trial, those who spoke of "Darwinism" were often referring to the mechanism of natural selection rather than the general idea of evolution.

Darwin began, cleverly, with a chapter on "Variation Under Domestication," safely focusing on the very familiar phenomenon of *artificial* selection. It was well established that humans had created particular types of plants and animals possessing desirable traits—for example, different breeds of dogs or (especially popular among hobbyists at the time) different breeds of pigeons that varied in plumage and adornment. Desirable traits could be identified and then promoted and emphasized through controlled breeding. In successive generations, this resulted in the accumulation of incremental changes, eventually modifying the original organism in dramatic ways. Anyone who has compared a Great Dane to a Chihuahua will be broadly familiar with the principle.

Darwin then considered variation under *natural*, rather than artificial, conditions. He argued that the concept of species was unclear, with no firm criteria as to what features should be deemed necessary or sufficient for species assignment. Darwin stressed differences among naturalists as to whether or not particular features conferred the status of a "variety" within its species, demonstrating that distinctions between "variety" and "species" were vague and arbitrary. He then clarified this apparent taxonomic chaos by proposing that similarities and dissimilarities arise as a result of descent with modification from a common ancestor.

Darwin's primary mechanism to account for the diversity of life forms was the process of natural selection. Individuals within a species differ in their morphology, physiology, and behavior: variation is universal. Some of this variation is heritable; on average, offspring tend to resemble their parents more than they resemble other individuals in the population. Organisms have a huge capacity for increase in numbers; they produce far more offspring than give rise to breeding individuals. Individuals compete for scarce resources such as food, water, and even mates. (Darwin's thinking here, like Wallace's, was influenced by the English economist and demographer Thomas Malthus, famous for the theory that human population growth eventually outpaces the food supply.) Darwin reasoned that because of competition, some variants will leave more offspring than others. These offspring inherit beneficial characteristics of their parents, resulting in evolutionary change taking place via "natural selection."

In consequence, at any given time, organisms tend to be adapted to their current environments. Unlike Chambers, Darwin realized that the process of adaptation does not necessarily result in more complex outcomes compared with ancestral forms. Instead, individuals who are better able to find food, water, and mates (while avoiding predators) will, on average, have greater representation in future generations (regardless of simplicity or complexity). Darwin pressed his point with a remarkable collection of observations about the manner in which organisms vary.

Darwin also devoted a large section of *Origin* to proactively address problems with his theory—issues he acknowledged and knew might trouble his readers. These potential concerns included clear gaps in the fossil record, which Darwin realized were unavoidable because fossils are not inevitably formed and are dependent upon specific conditions for their preservation. He also discussed how complex adaptations, like the eye, could have evolved from earlier stages that would not involve their present complexity. To this possible objection, he responded with evidence that simple eyes serve their owners in ways commensurate with the ecological demands of their existence. He thus noted that mollusks are one animal group that includes taxa with complex lens-based eyes (squid, octopus) and others with much simpler eyes (clams, mussels), along with a range of intermediate forms.

Another question that puzzled Darwin was the existence of extravagant traits, those that are apparently costly, rather than advantageous, to the individual. A classic example is the peacock's tail. To explain these, Darwin introduced the concept of "sexual selection" as a rider to the broader process of natural selection. Darwin proposed that some traits that were hard to conceive of as adaptive with respect to *survival* may nonetheless be adaptive because they conferred a mating advantage, in that the trait was attractive to the other sex.

Origin was written accessibly and even self-deferentially. Darwin had voluminous evidence on his side, and another author might have adopted a sanctimonious or imperious tone, but he was never bombastic or condescending. The result was a disruptive idea presented in an approachable manner.

But *Origin*'s cataclysmic impact resulted not just from Darwin's insights and style but also from what we might today label as his publicity campaign. Although nothing comparable to the launch of a modern bestseller existed at the time, Darwin's *Origin* was nonetheless a publishing event. After consulting with Lyell, Darwin chose publisher John Murray, who had published Lyell's own *Principles of Geology*. Murray had two trusted and experienced reviewers read Darwin's manuscript and decided to proceed (although one assessment was negative).

Now at age fifty, Darwin was transformed and consumed by his dedication to seeing that his work was given a proper reception and a fair assessment. Here, Darwin did what Wallace never could, benefiting greatly from the financial security that facilitated his full-time devotion. He launched a vigorous letter-writing campaign. Due twelve gratis copies from the publisher, Darwin purchased eighty more to send to carefully selected recipients. Some recipients were notables that Darwin reasoned would help his cause; others were those he deemed potential critics who might be nudged away from a negative reaction upon receiving a personal copy from the author. Darwin fine-tuned the accompanying letters with self-deprecation and flattery appropriate to the recipient—no simple "with compliments of the author" here—all with the purpose of bolstering support.

Not all recipients were won over. Darwin's former geology teacher, Adam Sedgwick, who had mentored him before the voyage of the *Beagle*, wrote to Darwin that he read the work "with more pain than pleasure," conceding that he admired some parts but judging others to be "utterly false & grievously mischievous."[19] (This critique notwithstanding, they remained friends.) But Darwin's letter-writing campaign succeeded in cultivating interest.

The unadorned 502-page book sold all 1,250 copies on its first day. Of course, new books by popular authors of the day, such as Dickens or Trollope, had much larger print runs. But selling out on day one was still an achievement, and Darwin and Murray were both pleased. Despite some initial fears, the subject matter was apparently not too dry to appeal to the general public. With six editions, the classic work has remained in print continuously from its publication date of November 24, 1859.[20]

But inside both the church and the academy, the personal prestige of many luminaries, their long-held worldviews, and their authoritative influence were now threatened—not so much by the idea of evolution (which, again, was not a novel one), but by the coordinated, radical, and specific arguments in *Origin* combined with the widespread attention it received.

Darwin declared, modestly but eloquently, in his most famous book's final paragraph that there was "grandeur in this view of life." Yet despite this characteristically humble and cautious phrasing, he was fully aware that his conceptual innovations would prove revolutionary and transcendent. For Darwin had not just demonstrated that species change; he had suggested that they do so in response to competitive pressure from other organisms. To many, this was a materialistic view of life, an account that, in the eyes of some, minimized the role of God.

Today, Darwin stands as one of the most influential, and most scrutinized, scientists ever. Among the vast literature studying him, one provocative thought experiment was proposed by Peter Bowler,[21] a prolific and impactful historian of evolutionary science. Bowler argues that Darwin was uniquely positioned to derive the idea of natural selection at the time when he did—and that natural selection, which presented evolution "in its most materialistic" and thereby most controversial form, magnified ensuing controversies. Bowler thus imagines how evolutionary theory might have developed more gradually, and less controversially, had Darwin been swept overboard and drowned on the voyage of the *Beagle*.[22] (In reality, the *Beagle* was nearly lost twice during failed attempts to cross the Strait of Magellan before FitzRoy managed to steer her through.)

Darwin's insights would surely have been captured by others in time, gradually ratcheting up the public's understanding of evolutionary processes. The cataclysmic impact that he produced through *Origin* might have been replaced with incremental development, exposure, acceptance, and sensitization, perhaps leading to fewer battles like *Scopes*. In Bowler's words:

> In our world, Darwinism became a kind of bogeyman, an image invoked to frighten the faithful by highlighting how easy it was for science to undermine faith. Without that

> symbol, even conservative religious thinkers would have had less reason to fear the threat posed by the general idea of evolution.[23]

Bowler precisely captures the impact that *Origin*'s publication had on the controversy that exploded in 1859 and reverberated through the ensuing decades and centuries.

Yet that controversy was magnified not just by Darwin's *idea* but also by his *person*. Other writers, including Patrick Matthew, had put forth ideas that mirrored Darwin's natural selection. After Darwin's fame grew with the success of the *Origin*, Matthew complained that he was the true discoverer of natural selection, having outlined the idea in an appendix of an obscure 1831 book with the dry title *On Naval Timber and Arboriculture*, unknown to Darwin in 1859. And, of course, if Darwin had not existed, Wallace's version of natural selection would likely have been published in time.

Wallace himself, however, fully understood the essential differences in the scope of his and Darwin's accounts. Writing about Darwin to Henry Walter Bates in 1860, he said:

> I could never have approached the completeness of his book, its vast accumulation of evidence, its overwhelming argument, and its admirable tone and spirit. I really feel thankful that it has not been left to me to give the theory to the world. Mr. Darwin has created a new science and a new philosophy . . . Never have such vast masses of widely scattered and hitherto quite unconnected facts been combined into a system and brought to bear upon the establishment of such a grand and new and simple philosophy.[24]

In short, many nineteenth-century figures, including Wallace and Matthew, could have (and did) introduce ideas about natural selection to informed readers of works on zoology. Others, like Chambers, advanced cruder and less scientific ideas about evolutionary change to the general public. But only Darwin, with his unique combination of insights, position, and circle of effective and influential allies, could have done both—and done so with such an abrupt impact as to be perceived as a threat to traditional hierarchies and ideas. It was thus the shock of *Origin*'s introduction and Darwin's specific theories, rather than the general idea of evolution, that set the stage for the coming battles exemplified by *Scopes*.

In the next chapter, we show how these battles played out, introducing the initial pushback Darwin received from scientists who were skeptical of his ideas. The impact of the initial reaction to *Origin* was complex. Even as evolution as a general principle became firmly established and influential, certain objections to Darwin's explanations would plague his defenders for years to come.

CHAPTER TWO

One Long Argument Interrupted

> [P]ractically all of the zoologists, botanists, and geologists of this country who have done any work . . . believe, as a matter of course, that evolution is a fact, but I doubt very much if any two of them agree as to the exact method by which evolution has been brought about.
>
> —Testimony of Maynard Metcalf, Trial of John Scopes, Day 4

OVER SEVEN HUNDRED PEOPLE crowded together, eager to witness two heavyweights joust over Darwin's controversial theory and its implications for Christianity. On one side was a man who would become known as "Darwin's bulldog."[1] In his own words, he had been "sharpening up [his] claws & beak in readiness."[2] On the other was a staunch defender of traditional religion; he too had been preparing, and the eager crowd was largely on his side. Its size so exceeded expectations that the event had been moved from its original venue.

The date was June 30, 1860, and though these events echo the more famous *Scopes* trial, they preceded it by sixty-five years. This duel took place at the annual meeting of the British Association for the Advancement of Science, held at the newly opened Oxford University (Natural History) Museum, and it presaged arguments to come.[3] Remarkably, the meeting was held less than a year after Darwin's publication of *Origin*.

Already, the controversy kindled by a single book burned bright and heated.

THE FIRST CLASH

Darwin's champion, Thomas Henry Huxley, was an esteemed zoologist and life-long iconoclast (Huxley had coined the term "agnostic"). His antagonist, Samuel Wilberforce, was the bishop of Oxford and thus enjoyed a home-field advantage. Wilberforce was a prominent national figure from a famous family; his father had played a notable role in the fight against slavery in Britain. The bishop vehemently opposed the idea that humans are the product of evolutionary change, believing humanity's divine creation in its present form to be integral to the Christian faith.

Wilberforce had found a powerful ally and debate coach in the distinguished comparative anatomist Sir Richard Owen, Britain's most prominent biologist and the superintendent of the Natural History Department of the British Museum. Among Owen's many contributions was his study of the fossils of extinct reptiles; in 1842, he christened the animals we know as *dinosaurs*, combining the Latin words for "terrible" and "reptile."[4] He played a pivotal role in acquainting the public with these wondrous creatures, working with sculptor and artist Benjamin Hawkins to create life-sized replicas in London. To celebrate the opening on New Year's Eve, 1853, twenty-one scientists were invited to a sumptuous eight-course dinner inside a replica of a giant iguanodon.[5] Hawkins delivered a short presentation on the creation of the sculptures, and Owen then displayed his confrontational personality by criticizing the anatomical details of Hawkins' replicas in front of the assembled guests.[6]

Even though Owen himself had once been chastised in the *Manchester Spectator* for his 1849 suggestion that humans, through some natural process, might have evolved from fish[7]—

again, evolution was very much "in the air" before *Origin*—he developed a profound distaste for Darwin's idea of natural selection. Although open to the idea that new species might result from an evolutionary process, Owen thought that Darwin had not provided adequate evidence of natural selection's ability to produce new species through the accumulation of small changes. He had scathingly (and anonymously) critiqued *Origin* in the *Edinburgh Review*, the same year as the Oxford debate, arguing that Darwin's observations failed "not merely to carry conviction, but to give a colour to the hypothesis."[8] Owen thus agreed to lend a hand to Bishop Wilberforce, who would play the role of Darwin's chief antagonist at the upcoming British Association's Saturday session.

The session was chaired by Darwin's former Cambridge professor and mentor, John Henslow. Formally, no specific symposium on evolution was scheduled for the day. But a British-born American scientist from New York University, John William Draper, was scheduled to present a paper with the descriptive title, "The Intellectual Development of Europe (considered with reference to the views of Mr. Darwin and others) that the progression of organisms is determined by law."

In this work, Draper applied a metaphor of Darwinian adaptation to human social and political systems. Draper's views were complex: Although the application of evolutionary theory to social systems recalls the malign applications of "social Darwinism" (which we discuss later), Draper was a progressive and often enlightened thinker. For example, contrary to the beliefs of many contemporary scientists, Draper had argued that "there is, in reality, no difference in human races," asserting that, "[s]tripped of exterior coverings, there is in every climate a common body and a common mind."[9] Although willing to entertain connections between biological evolution and social change, he was not an anti-egalitarian like many later social Darwinists.

Regardless, whatever the merits of Draper's presentation, the crowd viewed it as a prelude to the real entertainment, when Bishop Wilberforce would respond, officially to Draper but really to Darwin. Wilberforce briefly addressed Draper with some derisive observations. Then, as planned and with the benefit of his preparation by Owen, he pivoted to attack Darwin's theory.

Wilberforce's assault is notable for its effort to challenge natural selection from a *scientific* perspective—despite the fact that Wilberforce's motivation (like that of most of Darwin's other critics) arose from concern over natural selection's implications for religion, morality, and human uniqueness. Wilberforce defended the permanence of species, noting that the remains of animals, plants, and humans from the Egyptian catacombs (some four thousand years old) revealed no differences with current species. To Wilberforce, such evidence revealed an "irresistible tendency of organized beings" to assume an "unalterable character." These comments elicited applause from the audience.

Wilberforce then criticized Darwin's account of domestication (artificial selection) in the creation of the many and varied breeds of pigeons. The bishop argued that, even in the case of the strikingly different types of pigeons that breeders had created, as soon as the animals were set free, their descendants quickly returned to the original "permanent" forms. Wilberforce's critique that selected traits might blend back into the general population presaged later headaches for Darwin and his defenders.

But the debate's most famous moment (and its closest analogue to later witness stand theatrics between Darrow and Bryan) came when Wilberforce asked whether Huxley claimed descent from an ape through his grandmother or his grandfather. This was a breach of the forum's unwritten rules and a tactical error at a time when public values placed a premium on civility. Huxley responded that he would prefer an ape for an ancestor than a man who used his abilities to distort and ridicule scientific study. After vigorously defending the voluminous evidence set forth in *Origin,* Huxley was considered by many to have won the intellectual battle.

This traditional narrative of the 1860 Oxford debate, which centers on heated and contentious exchanges between Huxley and Wilberforce, has been both supported and questioned by recent scholarship. James C. Ungureanu, analyzing letters from Draper, suggests that the event may have been less confrontational than often portrayed. In his own correspondence, Draper highlighted a more respectful—even cordial atmosphere—where both sides presented their views with a degree of mutual respect.[10]

However, Richard England's research, drawing on a more detailed report from the *Oxford Chronicle,* presents a different picture, more consistent with the traditional account. England's findings reaffirm the long-held view of the debate as a dramatic and intense clash—much like *Scopes* would be, years later—marked by sharp exchanges and a sometimes unruly atmosphere.[11] Together, these perspectives illustrate the complex and multifaceted nature of the nineteenth-century historical record, while also reflecting polarized views on Darwin's theory of evolution.

Darwin himself was not present at the proceedings because he had fallen ill. Regardless, public argumentation was not his way. Many eminent persons attended, however. Robert Chambers (the secret author of *Vestiges*) spontaneously introduced himself to Draper.[12] In the crowd as well was Robert FitzRoy, former captain of the *Beagle,* who had since been raised to rear admiral and had even served as governor of New Zealand. A gifted meteorologist, FitzRoy had come to the annual meeting to present a paper on storms. But the devout officer appeared at the Huxley-Wilberforce debate, holding a Bible and expressing regret for his inadvertent role in facilitating the formulation of Darwin's theory. Darwin did not take this personally—years later he donated £100 to a fund to provide for FitzRoy's family after the aging admiral died (by suicide and in poverty), matching the largest donation that had been made by other contributors.[13]

But strong opposition from men like Wilberforce, Owen, and FitzRoy was only beginning. *Origin* had started an intellectual battle, and its ideas were in the crosshairs. Darwin knew his fate was, for the present, in the hands of others, but he began collecting every review he could, entering notes in his journal about any criticisms mentioned and ideas about how to try to address these concerns in future revisions of the *Origin*. Those challenges, and at times, Darwin's responses, helped lead to competing theories about evolutionary mechanisms that prevailed by 1925. Darwin's account of natural selection as the dominant mechanism for evolutionary change had been meticulous and comprehensive, but the young field of evolutionary biology would soon splinter into the murk of competing ideas that were eventually presented by experts at *Scopes*.[14]

THORNS IN DARWIN'S SIDE

A primary early challenge to Darwinism centered around the age of the Earth. Although Darwin did not try to integrate a specific timeframe into his theory of natural selection, he appreciated that evolutionary change of the sort he was proposing would require enormous amounts of time, acknowledging that "it may be objected, that time will not have sufficed for so great an amount of organic change, all changes having been effected very slowly through natural selection."[15] Thus, in the first edition of *Origin*, Darwin estimated that the Earth was vastly older than three hundred million years. On this topic, he drew directly from his friend Lyell, who had earlier argued that the Earth was indeed ancient and had changed gradually but steadily from its original molten state.

Following Lyell's lead, Darwin reasoned that geological formations such as the striking chalk cliffs of coastal southern England were built up and torn down only very gradually over time. The remains of countless planktonic organisms formed the white mud and, eventually, the massive chalk accumulations. These were in turn subjected to the erosive and abrasive forces of wave action and wind.

Darwin also pointed to the Weald in southeast England. This region consists of the eroded remains of a massive dome of layered Lower Cretaceous rock, weathered, carved, and dissected by the elements to expose sandstone ridges and clay valleys. Darwin's calculation of the time this might have taken—at least three hundred million years—was based on the measurable thickness of the sediments and an estimated average rate for coastal erosion. As an additional argument, he suggested to his readers that to appreciate the enormity of time involved in these transformations, they "watch the sea at work grinding down old rocks and making fresh sediment."

He was on the right track, but shortly after *Origin* was published, Darwin recanted these calculations—thus injecting more uncertainty into the developing evolutionary timeframe. Critical reviewers pointed out that some of his assumptions (for example, that the erosion was mostly due to marine action) were incorrect, and Darwin dropped the example in later editions. An even more threatening challenge came from a formidable opponent, Lord Kelvin (William Thomson, 1824–1907), discoverer of the first and second laws of thermodynamics, and a devoted Christian. Kelvin charged that the Earth (and even the Sun!) had not existed long enough for evolution to have taken place at the exceedingly slow pace Darwin imagined.

Kelvin used the rate of the cooling of the Earth from its original molten state to calculate that the planet was some hundred million years old—far too young for natural selection to work in the way Darwin had proposed. In his autobiography, Darwin acknowledged that as a student, he had "attempted mathematics," but that he had "got on very slowly." He and his supporters were hard-pressed to rebut Kelvin's authoritative challenge. Grudgingly, they instead responded to the new constraints of a much younger Earth with modifications to the original concept of natural selection. This was a step backward.

Even great physicists occasionally miscalculate, and Kelvin was wrong about the age of the Earth. Early assessments of how he erred suggested that his models did not include the effect of radioactivity (specifically its heat-generating properties), which had not yet been discovered. More recent scholarship highlighted that Kelvin's model was based exclusively on the principle of conduction, the transfer of heat energy by direct contact.[16] Notably, one of Kelvin's junior assistants, John Perry, proposed a different model at the time, but Perry had been ignored by the master. Perry's model was based on convection, the movement of heat by the physical motion of matter (in this case, continental drift), and it was the appropriate one. In time, Perry's model would restore an estimate closer to the one Darwin needed.

A separate dilemma, still very much alive in 1925, arose around fossils. If extant forms of life derived from ancestral forms, why was the fossil record so scant? With respect to the lack of fossils, especially from the earliest life forms, Darwin admitted that he could "give no satisfactory answer." He conceded that the "case at present must remain inexplicable; and may be truly urged as a valid argument about the view here entertained."[17] Indeed, the idea that "gaps" in the fossil record undermined evolution quickly entered the public consciousness. For example, an otherwise obscure Texas court decision in 1911 casually analogized between a faulty chain of logic and the fossil record's supposed lack of support for evolution.[18]

Darwin realized, however, that the formation of fossils requires particular and rare conditions. For example, fossils form when, shortly after death, an organism is quickly covered by sediments like mud, sand, or volcanic ash. These conditions are never universally present. Fossils are thus a rare gift, not a regularly scheduled event, and it is too much to hope for anything like an encyclopedic record of past lives. Nevertheless, the discovery of the feathered dinosaur *Archaeopteryx* and other key fossil finds gave Darwin confidence that this criticism of his theory would eventually wane as further discoveries were made. And indeed, by the time of *Scopes*, the fossil record was generally seen as supporting, rather than undermining, the idea of evolution.

A more challenging objection than supposed gaps in the fossil record originated from Fleeming Jenkin, a talented and successful engineer born not far from Darwin's home. Jenkin's achievements were notable in their own right: he collaborated with Darwin's other main tormentor, Lord Kelvin, to develop the first transatlantic cable, and his eventual biographer was famed author Robert Louis Stevenson.

Jenkin argued that any novel and advantageous trait that arose in a small number of individuals would be quickly diluted in a few generations—and thus lose its relative advantage—because of inevitable mating with individuals lacking the relevant trait.[19] According to Jenkin, the vastly greater number of individuals *lacking* the adaptive trait would doom it to insignificance in a few generations. How, then, could natural selection work?

In his review of *Origin* (written when he was twenty six), Jenkin illustrated his point with a disturbingly racist hypothetical.[20] Jenkin critiqued Darwin's theory by hypothesizing a shipwrecked white man on an island. Jenkin supposed the shipwrecked man to possess various superior traits "of a dominant white race" which would allow him to rule the island and its indigenous people. In Jenkin's thought experiment: "Our white's qualities would certainly tend very much to preserve him to good old age, and yet he would not suffice in any number of generations to turn his subjects' descendants white."[21] Scientific racism of this sort was not universal—as noted, Draper's work, quoted already, provides a counterexample—but it was nonetheless far from uncommon at the time among scientists on both sides of the evolution debate. It would be many years before the research community, which has incrementally incorporated greater diversity, began to grapple with this legacy of prejudice.

Decades later, in the late 1950s, scientists adopted a far less egregious analogy for the blending inheritance and swamping issue raised by Jenkin, labeling it "the paintpot problem."[22] They compared the novel trait to a can of red paint and imagined mixing it with white paint, representing the trait common in the population. The result would be pink paint, and there would be a further decline of red pigment with each additional mixing with white pigment.

In the fifth and sixth editions of *Origin* (and in later of his books), Darwin cited Jenkin's published critique, modifying—and weakening—several of his original thoughts and positions in an attempt to respond. By 1871, in the *Descent of Man*, Darwin concluded "that in the earlier editions of my *Origin of Species* I perhaps attributed too much to the action of natural selection or the survival of the fittest."[23] The alternative ideas Darwin employed in later works were ones he had entertained and dismissed years before. They include, for example, larger roles for abrupt organic change and quasi-Larmarckian[24] acquisition and transmission of traits. By walking back his original insights, Darwin left the field open for competing theories that would contribute to an impression of uncertainty. Darwin died with the blending inheritance problem unsettled.

The issue of heredity led to still more challenges to the Darwinian canon. How exactly do novel traits arise, and through what process are they transmitted to the next generation? Darwin and, as he was keenly aware, his critics all realized that evolutionary theory would never be complete without better knowledge of the mechanisms through which traits are passed down to offspring. In *Variation of Animals and Plants Under Domestication* (1868), Darwin outlined what he acknowledged was a "provisional" hereditary theory, which he referred to as pangenesis. He proposed that each part of the body—in fact, each cell—generated entities he termed "gemmules" that would be passed on to offspring by both parents and would control the development of the equivalent body parts of offspring. Gemmules, Darwin suggested, might travel through the bloodstream with the reproductive organs as their destination. Thus, in a quasi-Lamarckian manner, traits acquired by a parent could be passed on to offspring. Because both males and females supposedly contributed gemmules, particular characteristics in offspring might appear as a blend of the parents' traits.

For years, prior to 1859, Darwin had accepted both the soft and hard inheritance of traits.[25] By the publication of *Origin*, he clearly rejected the notion that major induced or inflicted bodily changes, such as mutilations, could be transmitted to offspring. But although he was vague about it in the *Origin*, Darwin clearly accepted soft inheritance in the form of the direct effect of the environment and, especially, the effect of "use and disuse."[26]

Darwin was wrong about pangenesis and the idea never really caught on, even among his core supporters.[27] By the turn of the century, Mendelian genetics began to provide key insights into inheritance. Even leading up to that, ideas about soft inheritance, like pangenesis, had been dealt a severe blow by the highly creative and influential German biologist, August Weismann (1834–1914). Weismann, an ardent Darwinian, himself appreciated that scientific controversies surrounding Darwinism would never be settled without a comprehensive understanding of inheritance.[28]

Weismann was responsible for many discoveries and proposals[29] that would bridge to modern evolutionary theory. These included, probably most famously, the conceptualization establishing the separation of the body (soma) and what he called the "germ plasm," later to be identified as genes.[30] Separation, here, refers to the hypothesis that although experiences and environmental factors can readily modify the body, these changes are not passed on through the hereditary material.[31]

Early in his career Weismann, and many others, accepted some form of the inheritance of acquired characteristics, especially the inherited effect of use and disuse. But Weismann's rejection of soft inheritance was clear and complete by 1883. He theorized that heredity occurs as a result of transmission from one generation to the next via some substance with definite chemical and molecular constitution, a brilliant insight for the time. This "hard" heredity left no opening for the generational transmission of acquired traits. Weismann's absolute rejection of soft inheritance antagonized some of Darwin's supporters, because of its important (if somewhat ambiguously endorsed) role in *Origin*, especially its later editions.[32]

Through Weismann, the inheritance of acquired traits lost currency.[33] But the so-called "Neo-Lamarckians"—a group of several prominent physicists, philosophers, writers, and biologists, including Trofim Lysenko (whose misguided ideas about inheritance proved catastrophic for Russian genetics and agriculture over a thirty-year period) persisted with the idea, arguing that environmentally induced behavioral modifications were critical to species change and development. They believed in evolution but decidedly rejected natural selection and the randomness of variation.[34] Darwin's protégé George John Romanes, in turn, coined the term "Neo-Darwinism" in 1905 to describe an opposing version of natural selection without any role for the inheritance of acquired characteristics.

The new century thus saw several contradictions. Evolution itself was, by now, not in question for most of the research community. But Darwin's theory of natural selection was in limbo, if not outright decline, as the primary proposed mechanism for evolutionary charge. For the wider lay audience, this created growing room for doubt. The inseparable branding of Darwin's name to the evolutionary paradigm had consequences. If the core ideas presented in the blockbuster *Origin* were falling out of favor—if Darwinism was failing in critical respects—what were the implications for the general acceptance of evolution? These questions and contradictions set the stage for arguments that would be even more heated than those between Huxley and Wilberforce.

In the following chapter, we advance further down the road to *Scopes* by tracing the outward migration of these ideas from Britain to America and from scientific fields into other disciplines, including law, economics, and social theory.

CHAPTER THREE

Survival

> [T]his loathsome application of evolution to social life . . . Can any Christian remain indifferent?
>
> —Planned Closing Argument of William Jennings Bryan, Trial of John Scopes

ALTHOUGH THE EVENTS RECOUNTED so far largely occurred in the United Kingdom, the impact of *Origin* was by no means limited to Darwin's home country. Its reception in the United States was far from uniform; the country was on the verge of the Civil War at the time, and many minds were thus focused elsewhere. But two prominent scientists, Asa Gray and Louis Agassiz, emerged as central figures in the debates surrounding evolution and Darwinian natural selection within US academia. And discussion of Darwin's ideas spread quickly to the general public.

DARWIN AND AMERICA

Gray, a renowned American botanist at Harvard, was 49 years old when *Origin* was published. He proved critical to the reception of evolutionary theory in the United States. More than anyone else in the era before *Scopes*, he was Darwin's American bulldog. The two were also firm friends: they corresponded extensively over the years, beginning in 1855, and Gray even traveled to Darwin's Down House for visits in 1868 and in 1881. Gray was thus a natural choice to author the introduction to the US edition of *Origin*.[1]

In America, Gray's endorsement of Darwin's ideas carried weight in both scientific and religious circles. His contributions were multifaceted and accommodating, in that Gray sought to reconcile evolution with his devout Christian faith. In his 1861 essay, *Natural Selection Not Inconsistent with Natural Theology*,[2] Gray defended the compatibility of evolution and religion. This approach, he believed, would ease tensions between religious beliefs and this new evolutionary view of life. Gray's most famous work, *Darwiniana*,[3] a collection of essays published in 1876, represented his clear commitment to Darwinism, as well as his efforts to reconcile evolution with his own devoted religious belief.

He did this, in part, by suggesting that variation in nature was essentially predetermined in ways that resulted in beneficial outcomes through natural selection. The implication was that God was responsible for, and directed, the laws of variation. Gray's efforts helped establish the credibility of Darwinism in the United States while softening its introduction; they influenced subsequent generations of both scientists and theologians in that his advocacy provided a platform for American scientists to engage with Darwinism without compromising their faith. With Gray serving as Darwin's primary American ambassador, it would have been difficult to predict, at the beginning, the intensity of the coming religious attacks on evolution in the United States.

But in stark contrast to Gray, Louis Agassiz, a 52-year-old Swiss-American naturalist and geologist, adamantly opposed Darwin's theory of natural selection. Agassiz founded the Harvard Museum of Comparative Zoology, and his professional reputation at the time surpassed that of Gray. He was not only the country's leading naturalist but also one of the world's leading authorities on fossil fishes. Agassiz strongly believed in the separate creation of species, content with the established notion—shared with his famous mentor, the French naturalist, zoologist, and anti-evolutionist Georges Cuvier—that each species was independently fashioned by a divine creator.

Although Agassiz's deeply rooted religious convictions formed the cornerstone of his scientific rejection of evolution, they did not preordain it, because Asa Gray was devout as well. Later scholars thus suggested that Agassiz's failure to embrace Darwin's novel evolutionary proposals resulted as much from his traditional training and conceptual framework, rooted in Cuvier's antiquated idealism,[4] as it did from faith. But whatever the motivation, Agassiz's status meant that his rejection of *Origin* carried significant weight within American academia, and he lectured and argued against Darwinism for years. Initially, he was remarkably unyielding in his stance, and his opposition to natural selection became legendary within academic circles. Within a decade or so, however, he softened his criticism to a degree, realizing that the Darwinians were gaining ground.

Fascinatingly, Darwin's core ideas, which he had shared with Gray before publishing *Origin* in September 1857, were the subject of a set of formal public debates between Gray and Agassiz beginning at the American Academy of Arts and Sciences in Boston—predating the far more famous clash in England between Huxley and Wilberforce.[5] These debates centered on the origin and nature of species, as evidenced by their geographical distributions. Gray presented his observations on the striking similarities of various species of Japanese plants with those found east of the Mississippi River in the United States and questioned the likelihood that "each species originated where it now occurs . . . at the present time," which was Agassiz's position. Gray's Darwinian argument was that species had originated at a single location, diversified, and disseminated to particular regions of the world, accounting for the observed similarities. The notes from the meeting documented Gray's argument that "the idea of the descent of all similar or conspecific individuals from a common stock is so natural, and so inevitably suggested by common observation, that it must needs be first tried upon the problem; and if the trial be satisfactory, its adoption would follow as a matter of course."[6]

Agassiz's response was brief and consistent with his long-held views on the separate creation of species, summarized in the notes as: "that distinct species were originally created about in the positions which they now occupy upon the Earth's surface . . . [I]n the Animal Kingdom facts point that way. There may be a difference between animals and vegetables in this respect."[7]

Anticipating Darwin's core argument, but interpreting it in a completely opposite manner, Agassiz went on to protest that the "warfare" that commonly goes on between species in competition over resources made it impossible to imagine that one species "could have been originated in a single pair." His firm mindset about the fixity of species led him to miss alternative interpretations. At one point in the proceedings, Agassiz employed what would become a common tactic in such debates: cast doubt on your opponent's knowledge and abilities: maybe, he suggested, the problem was that Gray's classification of the plants he had studied had been incorrect, and there were perhaps more differences among species than had been reported. Unsurprisingly, Gray and Agassiz also disagreed on the origins of the human species, with Agassiz believing that different human races had been separately created. Gray, in line with Darwin, believed in the shared evolutionary ancestry of all humans.[8]

The evolution debate had crossed the Atlantic—and took shape in ways that paralleled what happened back in England. But ironically (given developments to come), in the United States, debate first took the form of civil academic disagreements. Despite their intellectual differences, Gray and Agassiz remained cordial, and early arguments over evolution in America played out between scientists with different philosophical and academic views. Needless to say, that would change.

DEFENDING THE FAITH

Gray's efforts at reconciliation notwithstanding, the potential threat that evolution posed to traditional religion was not illusory. Studying at Harvard during this period (and at one point, enrolled in Gray's botany class[9]) was a young Oliver Wendell Holmes Jr., later to become perhaps the most influential jurist in American history.[10] Looking back on his youth and his drift away from religion, Holmes would later write:

> [M]y father was brought up scientifically—i.e. he studied medicine in France—and I was not. Yet there was with him as with the rest of his generation a certain softness of attitude toward the interstitial miracle . . . that I did not feel. The difference was in the air, although perhaps only the few of my time felt it. *The Origin of Species* I think came out while I was in college . . . I hadn't read [it] to be sure, but as I say it was in the air.[11]

By positing that the complexity of life arose from natural processes, Darwin had uprooted the traditional Western worldview that saw the intricacies and apparent design of living beings as evidence of a divine creator. Some American elites, like Holmes, reconciled to this new view. But to many, the safety and security of a static and divinely ordained natural world was a cause they would not easily abandon. Evolution was—and even today continues to be—perceived as a significant threat to established belief systems and time-honored doctrines that enshrined humans as holding an exalted position in the universe.

Facing this challenge in the final decades of the nineteenth century, a number of Christian leaders in America began targeting evolution as a threat to public morality and the Bible. Over the next forty years, bridging to the time of the *Scopes* trial, the major American Protestant denominations evolved in their views about science, with liberal modernist groups and the theologically conservative fundamentalists gradually coming to perceive the threat of evolution differently. The liberal denominations, following the lead of scientifically minded Christians like Gray, accommodated evolution into their belief systems and felt that their faith could coexist with evolutionary theory. This was much less so for the fundamentalists, for whom the literal accuracy of the Biblical text was a non-negotiable point of faith.[12]

Of course, many other scientific discoveries also stood in tension with absolute biblical literalism. Even Bryan, the fundamentalists' great champion at *Scopes*, would balk when asked to commit to the position that the universe had been created in just seven twenty-four-hour days. But for many in the fundamentalist movement, evolution represented a more comprehensive and existential challenge to the faith than had nineteenth-century discoveries from other fields. This challenge led modern philosopher Daniel Dennett to call evolution "Darwin's Dangerous Idea." As Dennett explains:

> Whenever Darwinism is the topic, the temperature rises, because more is at stake than just the empirical facts about how life on earth evolved, or the correct logic of the theory that accounts for those facts. One of the precious things that is at stake is a vision of what it means to ask, and answer, the question "Why?"[13]

For many people of faith in the United States, the answer to the question "Why?" was (and is) interwoven with humanity's unique relationship with God, and that relationship was seen to depend on human uniqueness and immutability. Bryan would summarize this objection succinctly in 1925, focusing on the textbook that Scopes taught from, which depicted humans in a taxonomic circle with other mammals. "What shall we say of the intelligence, not to say religion, of those who are so particular to distinguish between fishes and reptiles and birds, but put a man with an immortal soul in the same circle with the wolf, the hyena and the skunk?"[14] Those who rejected what they deemed to be a materialistic account of human origins formed the antievolution movement, which clashed with religious modernists (to say nothing of the nonreligious) for decades to come—and continues to wield influence today.

In the latter half of the nineteenth century, however, this antagonism was not a one-way street. Some intellectuals in the Victorian period grew increasingly alienated from traditional faith, becoming convinced that modern science and organized religion stood in irreconcilable tension. John Draper, the New York University (NYU) professor whose remarks at Oxford had been the prelude to the Huxley–Wilberforce debate, published one of his most influential works in 1874, titled *History of the Conflict Between Religion and Science*.[15] Draper's thesis, apparent from the book's title, postulated inherent opposition between science and religion. The so-called "conflict thesis" is now controversial[16] (and was contradicted by the work of many of Draper's contemporaries, like Asa Gray). At the time, Draper's book sold well and was influential in some intellectual circles. Thus, the view that religion could not be reconciled with science was not held by the fundamentalists alone.

Nor were the great controversies over evolution restricted to matters related to science and religion. Evolutionary ideas rapidly became imbued with economic and political significance. Indeed, the first reference to natural selection in a US legal case used the term to describe the *marketplace*, not the development of species. Rejecting arguments about an unlawful monopoly in 1872, Justice William Howe of the Supreme Court of Louisiana wrote, "We may pity the weaker merchant, but we can not mulct the stronger one in damages. The great law of 'natural selection' is something we can not repeal, and 'the fittest survive,' and always will."[17]

Howe's opinion evoked a grim and primal picture of economic competition. And indeed, Gilded Age America was a land of staggering inequality, driven by the vast accumulation of wealth by a small number of entrepreneurs in powerful industries such as oil, banking, and railroads. Some of these men found in biological evolution an analogy and justification for the ruthlessly competitive capitalist system in which they had made their fortunes. Andrew Carnegie, in particular, wrote of "the law of competition" in his *Gospel of Wealth*, arguing, "We cannot evade it; no substitutes for it have been found; and while the law may be sometimes hard for the individual, it is best for the race, because it insures the survival of the fittest in every department."[18]

That striking phrase—"survival of the fittest"—originated with the polymath Herbert Spencer.[19] Spencer influenced the development of sociology as a formal academic discipline and shaped debates about the role of government, social progress, and inequality during his time. A contemporary of Darwin, Spencer was heavily influenced by Lamarckian ideas about evolution and believed human progress to be driven by population pressure. In this, he was greatly influenced by Malthus, whose influential eighteenth-century writing had postulated that human population growth would inevitably outstrip food supply, leading to competition and struggle within and between human populations.

Malthus's ideas about competition and scarcity within human societies had also influenced both Darwin and Wallace. But despite these shared influences, it would be overly simplistic to describe Spencer's ideas as an outgrowth of Darwin's. Malthusian thought long predated nineteenth-century advancements in biology, and Spencer's ideas about evolution were heavily influenced by Lamarck (who, again, predated Darwin). Nevertheless, Spencer's phrase "survival of the fittest" quickly became associated with Darwinian evolution (and some Darwinians saw it as a better descriptive term than "natural selection"[20]). In turn, the phrase "social Darwinism" gradually came to describe intellectual positions that evaluated human society through the lens of competition over scarce resources that would lead to adaptation within and among human groups.[21]

Not all attempts to analogize between natural evolution and human society were simplistic defenses of ruthless individualism. As noted, for example, the views of Draper were more complex. Charles Francis Adams Jr., an early theoretician of business regulation (and a direct descendent of Presidents John and John Quincy Adams) was himself a devoted Spencerite who often used the phrase "survival of the fittest" to describe the proper mode of competition between railroads; he was not, however, an apologist for robber barons.[22] Another author to compare social development to Darwin's ideas, without invoking ideas that would today be labeled "social Darwinism," was Henry James Sumner Maine (who chose to publish his influential work *Ancient Law* with John Murray, shortly after Murray had worked with Darwin to publish *Origin*).[23] Nevertheless, Carnegie's passage neatly illustrates how the simplistic application of biological principles to social theory could provide rhetorical justification for laissez-faire ideology. If competition was inevitable, natural, and ultimately beneficial, why should society intervene to moderate its impact?

As of 1873, Holmes's writing reflected the influence of Darwinian thought on political analysis. Having left Harvard to serve as an officer in the Union army[24] before returning to earn a law degree, Holmes penned a law review commentary on a harsh sentence handed down in England against gas stokers who had gone on strike in London. Liberals had criticized the English law as "class legislation" directed against workers. Holmes countered that the "struggle for life" does "not stop in the ascending scale with the monkeys, but is equally the law of human existence. . . . Why should the greatest number be preferred" by legislation? "Why not the greatest good of the most intelligent and most highly developed?"[25] Holmes's relationship with these ideas was complex and would change with time—in 1873, he was still years away from serving as a judge, and one of his most famous dissents chastised the majority of the Supreme Court for constitutionalizing laissez-faire ideology. But his early writing, like the Darwinian language invoked by the Louisiana Supreme Court, demonstrates the extent to which US intellectual circles engaged with Darwinian analogies to society during the Victorian era.

Indeed, they initially did so perhaps even more than they engaged with Darwin's actual insights. Holmes returned to evolution again as a metaphor for the development of legal doctrine in a common law system, writing in 1899 that the development of precedent was "a lively example of the struggle for life among competing ideas, and of the ultimate victory and survival of the strongest."[26] Over time, this metaphor between biological evolution and the progression of case law in a common-law system became nearly ubiquitous.[27]

The defining legal case of the early twentieth century was the Supreme Court's 1905 decision in *Lochner v. New York*,[28] which struck down a New York state law regulating maximum working hours for bakers. A bakery owner had been charged with unlawfully requiring an employee to work more than sixty hours in one week. But the high court invalidated New York's hours cap as unconstitutional, holding that the Constitution's guarantee that liberty may not be deprived without due process of law (located in the Reconstruction-era Fourteenth Amendment) enshrined a right to freely enter into contracts. Under this reasoning, an employee had a right to contract to work for more than sixty hours, and an employer had a right to bargain for that contract. A state legislature could not constitutionally interfere in the resulting deal.

This decision embodied a muscular and conservative style of jurisprudence, one that was ascendant, if not universally held, within the legal field. Contrarians argued that judges should defer to legislatures—which would entail upholding popular legislation like the regulation in *Lochner* (and, perhaps, like the state anti-evolution laws that were still a few decades away).[29] Holmes, by this point elevated to the highest court, advocated for deference.

Holmes came from an illustrious New England family and, personally, was anything but a progressive. Indeed, he bluntly declared in a private letter, "As to the *right* of citizens to support and education I don't see it . . . I see no right in my neighbor to share my bread."[30] But personal views notwithstanding, his understanding of the Constitution led him to defer to democratic legislation—progressive or otherwise—at a time when judicial nullification of reform was a relatively frequent occurrence.[31]

In *Lochner*, Holmes dissented vigorously and eloquently, highlighting the way that the Court majority's economic philosophy seeped into its legal reasoning, thereby constitutionalizing laissez-faire economics. Holmes explained that the case was

> decided upon an economic theory which a large part of the country does not entertain. If it were a question whether I agreed with that theory, I should desire to study it further and long before making up my mind. But I do not conceive that to be my duty, because I strongly believe that my agreement or disagreement has nothing to do with the right of a majority to embody their opinions in law.

He then noted dryly that the "14th Amendment does not enact Mr. Herbert Spencer's *Social Statics*," implicitly connecting the majority opinion with social Darwinism. But although the verdict of history is with Holmes, at the time he did not command a majority. Laissez faire reigned supreme, and many social Darwinists preferred it that way.

Contemporary scientists chafed against the use of biological principles to defend regressive legal, political, and economic outcomes. Botanist Lester Frank Ward, for example, objected that "I have never seen any distinctively Darwinian principle appealed to in the discussions of 'Social Darwinism.'" He contended that it was "wholly inappropriate to characterize as social Darwinism the laissez-faire doctrine of political economists."[32] Nevertheless, as the growing progressive movement began to challenge the economic consensus of the *Lochner* era, some saw an obstacle in evolutionary theory. For a young William Jennings Bryan, both a devout Christian and a strong progressive, the path to *Scopes* was already clear.

In the following chapter, we introduce Darrow and Bryan at the beginning of their stories, tracing how their paths brought them from an early alliance between two progressive crusaders toward a bitter rivalry over Darwin's ideas.

CHAPTER FOUR

The Toilers Everywhere

> We have the purpose of preventing bigots and ignoramuses from controlling the education of the United States.
>
> —Argument of Clarence Darrow, Trial of John Scopes, Day 7

DARWIN'S GENERATION SAW PROMINENT public debates over his ideas, including Huxley against Wilberforce in England and Gray against Agassiz in America. But it was the next generation's similar clash—in a courtroom, rather than an academic forum—that became the enduring symbol of debate over evolution. The pioneering ACLU attorney Arthur Garfield Hays, himself a lawyer at *Scopes*, later described the duel between Clarence Darrow and William Jennings Bryan in broad terms that echoed Draper's conflict thesis: "a battle between two types of mind—the rigid, orthodox, accepting, unyielding, narrow, conventional mind, and the broad, liberal, critical, cynical, skeptical and tolerant mind."[1]

But if their mindsets were truly so diametrically opposed, the difference was not so apparent until the latter stages of their careers. Long before they became antagonists, the two men were fellow travelers, with Darrow a crusading attorney for the labor movement and Bryan a fierce champion of progressivism. That they ended so far apart is a testament to the divisive cultural conflict surrounding evolutionary theory, as well as to the entrenched features of the US legal, economic, and political systems that ultimately drove both men to redirect their lives. Our account thus traces the public careers of these two men who did so much to cement the legacy of the *Scopes* trial.

FELLOW TRAVELERS

Clarence Darrow came to movement advocacy as a young man, influenced by the progressive politician John Peter Altgeld's book *Our Penal Machinery and Its Victims*.[2] Like Darwin long before him, Altgeld launched his book with a publicity strategy, buying and sending extra copies of his passionate critique of over-incarceration to influential figures across the Midwest. One copy found its way to Darrow, a young attorney then enroute to Chicago. Darrow was immediately attracted to its critique of repressive, Gilded Age power structures embedded in criminal law.[3] Altgeld, a Democratic power broker in Darrow's adopted city, took the promising young attorney under his wing. Darrow's biographer, Andrew Kersten, wrote that Altgeld became "as important a figure in Darrow's life as his parents."[4]

Darrow argued his first epochal case before the Supreme Court in 1895, establishing his reputation as an effective attorney willing to represent labor radicals. The case arose from the Pullman Strike, a massive labor action initiated by the American Railway Union to support workers locked out of a Pullman rail-car manufacturing plant. The strike consumed Chicago and shut down most rail traffic in the western half of the country. By then, Altgeld was Illinois's governor, but he could do little when President Grover Cleveland deployed the army to his state and city. Contemporary observers likened the resulting strife to the French Revolution.

In the midst of the chaos, a federal court enjoined the union and its leader, Eugene Debs, ordering the strike dissolved. Debs refused and was arrested, charged with conspiracy and contempt for violating the court's injunction.[5] Darrow took the case. Defending Debs with rhetorical flourish, Darrow won over the jury in the conspiracy trial. But Debs's violation of a federal court's injunction was a separate legal matter—one that would be adjudicated by a judge, not a jury. Darrow brought that case to the nation's highest Court, arguing that the injunction against the strike was a perversion of the Sherman Antitrust Act (a law originally designed to curtail corporate power, now turned against workers by Cleveland's Attorney General). The Supreme Court was unmoved, and Debs went off to jail—not for the last time.

In re Debs was one of a trio of Supreme Court decisions in 1895 that prioritized business interests over labor and property rights over popular legislation. The other two invalidated an income tax and exempted a sugar monopoly from federal antitrust regulation.[6] Although the Court was not monolithic in its treatment of progressive reforms, and most Americans continued to hold the Court in high regard, decisions like these contributed to expanding popular resentment against a conservative judiciary. Reform advocates charged that unelected judges blocked majorities from enacting legitimate reforms while using injunctions to curtail labor activism without the involvement of juries.[7]

These contested decisions came down during the lead-up to an election year when Democratic party politics were largely consumed by debate between supporters of the unitary gold standard (including business interests and Grover Cleveland) and supporters of gold-and-silver bimetallism (which appealed to populists). Darrow's patron, Governor Altgeld, was German-born and could not run for president, but he hoped to play kingmaker at the Democratic national convention, which would be held in his home state. As a strong progressive, Altgeld aligned with the bimetallists. Darrow helped his mentor on the campaign trail, delivering speeches against the gold standard and critiquing the argument that economic necessity mandated that the United States match British commercial policy. In keeping with the age's heated rhetoric around currency issues, Darrow compared America waiting for Britain to consent to bimetallism with "slaves in the south waiting for the masters to consent to freedom."[8]

Darrow came to the convention with Altgeld, who pushed for the nomination of a bimetallist Missouri senator, Richard "Silver Dick" Bland.[9] But Bland was overshadowed by a younger bimetallist from Nebraska: William Jennings Bryan. Standing before the convention, Bryan decried the gold standard in a fiery speech that won him the nomination and thrust him onto the national stage:

> Having behind us the producing masses of this nation and the world, supported by the commercial interests, the laboring interests, and the toilers everywhere, we will answer their demand for a gold standard by saying to them: "You shall not press down upon the brow of labor this crown of thorns; you shall not crucify mankind upon a cross of gold!"

Bryan's skill in blending calls for economic redistribution with the evocation of Christian morality set him up for a long career as an effective speaker. Wild enthusiasm ensued, and Bryan became the youngest-ever major-party candidate for president.[10]

Not all were taken in. The young liberal lawyer Louis Brandeis (who would later jointly represent a client with Darrow,[11] in a brilliant career that would eventually earn him a seat on the Supreme Court), wrote of Bryan's speech: "It must have been the voice and the manner, and above all the general temper of the audience that carried all away. There is general sadness in the situation."[12] Similarly disappointed, Darrow would recall meeting with Altgeld the next day. The governor told his young protégé, "I have been thinking over Bryan's speech. What did he say, anyhow?"[13]

But notwithstanding these dissenters, Democratic fervor for Bryan's campaign was real. In keeping with the populist zeitgeist, Bryan took aim at the Court as well as currency policy. To be sure, silver played a far larger role in Bryan's campaign than the right to strike. But under his leadership, the Democratic platform decried "government by injunction as a new and highly dangerous form of oppression by which Federal Judges . . . become at once legislators, judges and executioners," a direct reference to cases like the recent ruling against Darrow and his client Debs.[14]

But it was not to be. Bryan lost to McKinley, Altgeld lost to his gubernatorial opponent, and silver lost to gold. The disappointed Brandeis would again write, "I forgive the great West its Bryanism—and all the vagaries economic and social which that broad term comprises."[15] But neither Bryan nor "Bryanism" were done. The stirring speaker, who would be dubbed "the Great Commoner," ran on the Democratic ticket against McKinley again in 1900. He met with no more success in the rematch.

Shortly thereafter, Bryan and Darrow corresponded with each other, united by mutual admiration for Altgeld, who died in 1902. Bryan helped Darrow raise funds to support Altgeld's widow, and both men spoke at the funeral service.[16] For now, their paths crossed amicably.

REFORM AND INTRANSIGENCE

A reformer *did* soon ascend to the presidency: McKinley's assassination in 1901 elevated Theodore Roosevelt, who among other things, battled monopolies and trusts. Roosevelt's character and philosophy, however, were quite different from Bryan's. Their differences were not limited to party or policy. Roosevelt took an eager interest in evolutionary theory throughout his life. In 1916, he would author a *National Geographic Magazine* article in which he expressed deep interest in "the crucial period in the evolution of man from a strong and cunning brute into a being having dominion over all brutes and kinship with worlds lying outside and beyond our own."[17] For now, he set about reforming and modernizing the American state.

But regardless of who held the White House, the *Lochner*-era judiciary remained intensely conservative and an implacable obstacle to reform. To be sure, the Court did not always reject reform in deference to property interests. One of several exceptions arose in Dayton, Tennessee, years before the small town drew the world's attention by putting John Scopes on trial. Dayton had experienced a minor boom when British investors organized the Dayton Coal & Iron Company to manufacture pig iron and coke. The firm employed around six hundred workers in its furnaces, mines, and coke ovens, paying up to $20,000 a year—although if workers wanted their pay at the end of a workday, they were forced to accept payment in company scrip, redeemable only at the company store.[18] Payment in scrip was outlawed by the state legislature in 1899, and a local merchant operating a competing store sued the company in the Rhea County courthouse. The case eventually reached the Supreme Court, and a majority of the justices upheld the reform against the interests of Dayton's most prominent business.[19]

But occasional victories like the Dayton case did little to convince progressive and labor-movement critics that the courts were on their side. The *Lochner*-era Court struck down maximum-hour laws, minimum-wage laws, child-labor laws, consumer-protection laws, and other reforms that had been hard-won through the democratic process.[20] State courts followed suit; infamously, the New York Court of Appeals struck down a new workers' compensation law as "plainly revolutionary," holding that the act violated the Fourteenth Amendment.[21]

The following day, over a hundred workers died tragically in a fire at the Triangle Shirtwaist Company, with poor working conditions contributing to the disaster and emphasizing the need for regulations.

The Fourteenth Amendment was a frequent point of contention. Originally enacted to safeguard the civil rights of freed slaves after the Civil War, it guaranteed in part that no *state* may deprive a person of "life, liberty, or property" without "due process of law." Under prevailing doctrine at the time, the first ten amendments—the Bill of Rights—were understood to apply only to the federal government, thus limiting their scope. But the Fourteenth Amendment's due process clause opened *state* laws up to federal constitutional challenges—and its language was vague enough to encompass multiple interpretations, including *Lochner*'s. Critics argued that the amendment no longer protected blacks, as was originally intended, but had instead become "the protector of all forms of commercial enterprises in all kinds of industrial pursuits."[22] Some progressives called for its repeal.

Other reformers, including Bryan, placed their hope in the popular election of judges, deeming that practice to be "the system toward which we shall approach as confidence in the stability of popular government increases."[23] With characteristic rhetorical flourish, Bryan noted that "the people are much more apt to deal justly with judges than they are to receive justice at the hands of judges who distrust the intelligence and the good intent of the masses."[24] Another crusading reformer, Senator Robert La Follette of Wisconsin, railed against the federal judiciary (particularly after the courts struck down laws prohibiting child labor), arguing for amendments that would allow Congress to immediately reinstate any laws that the courts had found unconstitutional.

But reforms intended to alter the appointment process for the federal judiciary or the power of judicial review required a constitutional amendment. At the time, the amendment process was not as insurmountable as it seems today: Over the course of his active career, Bryan campaigned for no fewer than four successful constitutional amendments (authorizing an income tax, the direct election of senators, prohibition, and women's suffrage). Even so, progressive amendments to reform the federal judiciary or repeal the Fourteenth Amendment remained well out of reach.

A more modest reform approach—and one that presaged the tactics used at *Scopes* and subsequent evolution cases—was judicial education, an attempt to integrate law and science. Within the bounds of legal theory, the incorporation of arguments from other fields was a controversial idea. The dominant jurisprudential style at the time viewed the law as an autonomous, internally valid, and closed system. This meant that legal questions were governed by internal principles found within the law itself, and these questions could be answered through syllogistic logic.[25] Although many turn-of-the-century legal thinkers saw law as its own kind of science, their conception of law as a complete and internally ordered system left little room for argument that incorporated discoveries and insights from *other* fields.[26]

But proponents of judicial education argued that judges should consider social and scientific evidence, regardless of whether that evidence was formally extraneous. Some judges were receptive; here, as in many areas, Holmes stood out. He believed the law could be improved through rationalization, proposing that legal decision makers should respond to scientific discoveries.[27] At times, he turned to evolutionary theory for analogy, attacking appeals to tradition for tradition's sake by arguing that, "just as the clavicle in the cat only tells of the existence of some earlier creature to which a collar-bone was useful, precedents survive in the law long after the use they once served is at an end."[28] He went so far as envision "an ultimate dependence upon science" by the law because science could allow decision makers to precisely determine the relative value of different social goals and thus choose optimal legal outcomes.[29]

Reformers hoped that open-minded judges might be persuaded to uphold popular legislation through the presentation of data, science, and analysis. A noteworthy success came in *Muller v. Oregon* when an Oregon state law fixing maximum hours that women could work in factories was challenged in 1908 by business interests.[30] Apart from the gender restriction, the case presented more or less the same issues that *Lochner* had. In defense of the statute, Louis Brandeis filed a famous brief compiling medical statistics from over ninety institutions. Although not formally relevant to the dispute between the parties, Brandeis marshaled statistics to show that long working hours were especially dangerous for women. In contrast to the result in *Lochner*, Oregon's statute was upheld. Legal briefs compiling and explaining data from experts in other fields have been referred to as "Brandeis briefs" ever since.[31] Here, the idea that legal argument might be supplemented and strengthened by reference to scientific developments found its root.

GENETICS, EUGENICS, AND CONTINUED ATTACKS ON DARWINISM

Although these efforts began to build bridges between science and law in the early decades of the twentieth century, some scientific fields were in a state of flux, their foundations unsettled. In particular, decades after the publication of *Origin*, the Darwinian evolutionary account remained under siege due to lingering problems, especially ones concerning the heredity of adaptive traits (specifically, their origin, stability over generations, and mode of transmission) as well as continued debate over the time frame required for natural selection to work. At the same time, questions related to heredity and genetics bedeviled the legal system as well, filtering through to advocates like Bryan and Darrow.

Key to these developments was work performed long before either man was born. In the nineteenth century, Gregor Mendel (1822–1884) conducted studies that would prove foundational to the study of genetics, heredity, and (eventually) evolution. Indeed, today, some claim that evolutionary theory's foundation depends as much on Mendel as it does on Darwin.[32] In some of his most transformative work, between 1856 and 1863, Mendel experimented with garden peas to investigate how individual traits are inherited. Although brilliant, his experiments were overlooked at the time.[33]

His pioneering research focused on seven distinct characteristics of pea plants: plant height, pod shape and color, seed shape and color, and flower position and color. Mendel's great insight was to focus on single traits at a time—others had attempted to examine heredity through breeding experiments, but they had simultaneously explored a jumble of traits, with confusing results. In one major experiment, Mendel crossed true-breeding yellow-pea plants with true-breeding green-pea plants. The offspring consistently produced yellow seeds. However, in the subsequent generation, green seeds reemerged in a ratio[34] of three yellow to one green. To account for this intriguing phenomenon, Mendel used the terms "recessive" and "dominant" to describe certain traits. The green trait, absent in the first filial generation, was labeled as recessive, whereas the yellow trait was identified as dominant.

Although initially overlooked, Mendel's studies elucidated how differences in imperceptible "factors" (now recognized as genes) were reliably associated with different traits. He concluded that alternative forms of factors (now called alleles) account for variations in inherited characteristics. For example, the gene for flower color in pea plants exists in two forms, one for purple and the other for white. For each trait, the plant (or organism) inherits two alleles, one from each parent. These alleles may be the same or different.[35]

Mendel hypothesized that allele pairs separate randomly, or segregate, from each other during the production of the gametes in the seed plant (egg cell) and the pollen plant (sperm). Because the process of meiosis separates pairs of alleles during gamete production, sperm or eggs carry only one allele for each inherited trait. When sperm and egg unite at fertilization, each contributes its allele, restoring the paired condition in the offspring. Mendel also found that each pair of alleles segregates independently of the other pairs of alleles during gamete formation. A "Mendelian trait" adheres to these principles of inheritance, characterized by its association with a single locus (gene) where alleles exhibit dominance or recessiveness.[36]

Around the turn of the twentieth century, Mendelian inheritance was rediscovered and recognized as transformational for understanding heredity and evolution. These insights are typically explained through three fundamental principles. The first is dominance: when an individual possesses both a dominant gene and a recessive gene for a particular trait, the dominant gene's expression will be observed in the phenotype. The second is segregation: in diploid organisms, where genetic information comes from both parents, the maternal and paternal alleles segregate randomly during the formation of gametes (egg and sperm), ensuring genetic diversity in offspring. The third is independent assortment: genes responsible for different traits are inherited independently of each other, meaning the inheritance of one trait does not influence the inheritance of another. This principle, too, contributes to the diversity of traits observed in offspring.[37]

After Mendel's studies of inheritance were rediscovered, innovative research in the new field of genetics—in particular laboratory work with fruit flies—gave rise to new perspectives on evolution. The tradition of understanding evolution through natural history, with its cataloging of specimens and observational approach, was ultimately supplanted by laboratory work.[38]

In principle, these efforts should have helped rescue the Darwinians and complemented the evidence for natural selection presented in *Origin*. (For instance, the Mendelian focus on "particulate" inheritance, namely, that parental characteristics do not inevitably blend in offspring, but instead are transmitted as discrete entities, addressed the core of Jenkin's dilution, or "paintpot" criticism). However, the new discoveries were instead hijacked to promote different evolutionary accounts, which attracted interest and gained prominence in the run-up to *Scopes*.

Natural selection, as Darwin and Wallace had conceived of it, was rooted in population biology: variation among individuals of a large population was continuous, and evolutionary change was gradual. This thinking was adopted by Darwin's cousin Francis Galton and his followers, the "biometricians," who developed and formally applied statistical approaches to the study of the inheritance of traits with continuous variation, like human height.[39] With the emergence of Mendelian classical genetics, however, many traits were recognized as discrete and discontinuous, and they appeared in predictable ratios (e.g., round pea versus wrinkled pea hybrid crosses producing a 3:1 ratio in favor of the round phenotype in progeny) with no intermediates. For the early Mendelians, this had anti-Darwinian implications in that change was categorical and sharp rather than continuous and slow. Over some thirty years, these two camps engaged in vigorous and sometimes rancorous argument.[40]

One anti-Darwinian alternative was "mutation theory," the idea that novel traits, and even new species, might appear in a more sudden ("saltationist") manner. This paradigm contended that mutation alone was the major mechanism of evolutionary change because it was the only known source of novel traits.

This basic idea of rapid and major transformation of species was not new[41] and Darwin had toyed with it himself. But he came to vigorously discount notions of abrupt change in the first edition of the *Origin*, writing *Natura non facit saltum* (nature does nothing in jumps).[42] In the early twentieth century, the idea was back in vogue, with prominent champions including Hugo de Vries, William Bateson, and T. H. Morgan.

De Vries took an especially aggressive anti-Darwinian approach. A Dutch geneticist, he[43] had independently rediscovered Mendel's laws, which had to that point been largely ignored. De Vries's studies of wild varieties of the evening primrose suggested that plants could differ significantly in various prominent features from parental generations, with back-crosses to the parental types sometimes impossible. He contended that new species might thus come into existence *de novo*, without the need for the gradual change that underpins Darwinian natural selection.

Mutation theory was attractive to many who had not been convinced by Darwin's argument that fossil formation is too rare to present a complete record. De Vries and other mutationists emphasized the evolutionary creativity of spontaneous discontinuity, arguing that it largely accounted for the origin of new species. To some, this spontaneity explained what they saw as gaps in the fossil record. Additionally, a mechanism for more rapid and significant change was seen as a way to address that persistent thorn in the side of the Darwinians: questions about whether contemporary estimates of the Earth's age corresponded to the lengthy time frame required for natural selection to operate. De Vries received international attention and went on numerous speaking tours, championing this new account of evolution. Withering under this attack, which focused on questions Darwin himself had acknowledged, Darwinian natural selection was becoming relegated to a historical footnote.[44]

De Vries was not alone in challenging Darwinian evolution. The geneticist Richard Goldschmidt (1878–1958) also argued that natural selection, with its emphasis on gradual change, could not explain how species originated. He sought possible genetic mechanisms that would allow for much more rapid change within phylogenetic lineages and suggested that change in a small number of critical genes, ones that regulated or controlled the expression of other genes, might be a mechanism to account for radically novel forms (so-called "hopeful monsters") and rapid evolutionary change. The thinking was that novel forms encountering new and different environments might, on occasion, be successful very quickly.[45]

Beyond sparking new challenges to Darwinian theory, the burgeoning field of genetics had other negative consequences, namely, the birth of eugenics.[46] The field traces back to Darwin's younger half-cousin, Francis Galton[47] (1822–1911), who was gifted in some ways not shared by his more famous cousin: quantitative thinking and analysis. In 1869, ten years after *Origin*, he published a book[48] analyzing the hereditary basis for intelligence. He documented high intellectual achievement in several prominent family lines (including his own family) and presented human intelligence in novel statistical terms, as a normally distributed trait with the tails of the curve representing small subsets of the population, with those intellectually gifted at one end and those inheriting much less-than-typical intelligence at the other.

Galton then suggested that humans, unique among living organisms, had the capacity to control their species' destiny by making decisions, for the good of the species, about which individuals should produce offspring (he believed the answer was those genetically gifted with high intelligence). He coined the term "eugenics" in 1883.[49] In his autobiography, Galton said the following about eugenics:

> Its first object is to check the birth-rate of the Unfit, . . . doomed in large numbers to perish prematurely. The second object is the improvement of the race by furthering the productivity of the Fit by early marriages and healthful rearing of their children. Natural Selection rests upon excessive production and wholesale destruction; Eugenics on bringing no more individuals into the world than can be properly cared for, and those only of the best stock.[50]

The idea caught on rapidly and at its peak, in the words of one scholar, was "supported by the entire range of opinion from the far left to the far right."[51]

The eugenics movement gained considerable traction, especially as it migrated to the United States from Britain. One of the first traces of this migration came in the writing of Oliver Wendell Holmes Sr., dean of Harvard Medical School and father to the pathbreaking judge, who wrote in 1875 that, if "genius and talent are inherited, as Mr. Galton has so conclusively shown; if honesty and virtue are heirlooms in certain families . . . why should not deep-rooted moral defects and obliquities show themselves, in the descendants of moral monsters?"[52] In the first decades of the twentieth century, legislative eugenics programs succeeded in the United States, where they had not in Britain. Advocates argued for policies to promote and encourage individuals with desirable traits to have children and to prevent or discourage reproduction by individuals considered "undesirable" based on their undesirable and deleterious traits.

The inaugural International Eugenics Congress, held in London in 1912, was chaired by Charles Darwin's son, Leonard, who had become an ardent supporter of the movement. By 1914, course listings for eugenics had appeared at Harvard, Columbia, Cornell, and Brown Universities.[53] In 1921, the second

International Eugenics Congress convened at the American Museum of Natural History in New York and was chaired by Henry Fairfield Osborn, with Alexander Graham Bell serving as honorary president. Content about eugenics was already present in high school curricula and appeared in many textbooks, including George William Hunter's *Civic Biology* (1914), the textbook used in the class taught by John Scopes.

But the ambitions of the eugenicists had outstripped the scientific evidence supporting them. *Civic Biology* included unquestioned support for eugenics:

> If the stock of domesticated animals can be improved, it is not unfair to ask if the health and vigor of the future generations of men and women on the earth might be improved by applying to them the laws of selection.[54]

The textbook's grounds for eugenics included summaries of the "notorious" (Hunter's word) examples of the "Jukes" and "Kallikak" (pseudonyms) family lines. Details of these families came from error-filled case studies used to support eugenic arguments in the late nineteenth and early twentieth centuries. The Kallikak account had recently been published by Henry Goddard, an influential and historically controversial psychologist, and purportedly illustrated the significant generational negative impact of hereditary traits associated with intellectual disabilities and criminal behavior.[55] Goddard went on to claim:

> Just as certain animals or plants become parasitic on other plants or animals, these families have become parasitic on society. . . . If such people were lower animals, we would probably kill them off to prevent them from spreading. Humanity will not allow this, but we do have the remedy of separating the sexes in asylums or other places and in various ways preventing intermarriage and the possibilities of perpetuating such a low and degenerate race.

In this context, *Buck v. Bell* reached the federal courts in 1927, challenging a Virginia law that provided for the compulsory sterilization of inmates in an institution for the "feeble minded." The Jukes family account was introduced as evidence.[56] When the case reached the Supreme Court, in the worst stain on his legacy, Holmes authored a nearly unanimous opinion (joined by Brandeis and others) upholding the law. The opinion noted that "experience has shown that heredity plays an important part in the transmission of insanity, imbecility, etc" and darkly stated, "Three generations of imbeciles are enough."[57]

The early-century connections forming between science and law—and Holmes's willingness to defer to novel legislation enacted by the political process—were thus far from universally positive. (As noted, Holmes's father had taken an interest in Galton's work as early as the 1870s. But *Buck v. Bell* was consistent with the jurisprudence of the younger Holmes and was joined by nearly all of his colleagues; it need not be explained as the product of family interest.)

Bryan and Darrow both opposed eugenics, but they approached the subject from different intellectual perspectives, presaging their transition from pro-reform allies to culture-war rivals. For Bryan, society's materialistic embrace of novel scientific theories was fundamentally misguided, and eugenics was a natural and inevitable outgrowth of the dangerous new doctrine of evolution. Bryan directly associated the rise of evolutionary theory with abhorrent practices like mandatory sterilization. In his view, supporters of evolution had offered society a program of "scientific breeding, a system under which a few supposedly superior intellects, self-appointed, would direct the mating and the movements of the mass of mankind."[58] Bryan saw a direct connection between the materialistic implications of evolution and moral decay.

In contrast, Darrow fully embraced science's inclusion in society and law. His friend Arthur Garfield Hays recognized that, to most lawyers, "science seems disassociated from their profession." Darrow, in contrast, had "devoted himself largely to the study of science, ranging through biology, sociology, and psychology." As an advocate, he used appeals to science to cast a defendant's actions as the product of "the inevitable processes of the universe."[59]

Like Bryan, Darrow opposed eugenics in part for moral reasons. He remained influenced by Altgeld's critique of incarceration, and he pointed out that there were far more humane ways to address problems like endemic crime than eugenics. But *unlike* Bryan, Darrow also approached the problem analytically. Though a scientific amateur, Darrow accurately explained that although it

> has grown commonplace in the discussion of crime to speak of isolation and sterilization as the proper treatment of the criminal and defective[,] [t]his is generally done without any clear understanding of the laws of heredity. The laws of the transmission from parent to child of traits and tendencies are not yet well enough known [. . .] There is much to learn, much to explain about the mysterious workings of heredity, before man can undertake to say that he has the wisdom or justice to choose the ones who should be the bearers of life to the future.[60]

Here, Darrow was absolutely correct. (Indeed, the absence of a full understanding of the "mysterious workings of heredity"[61] explains precisely why Darwin had been unable to effectively respond to the paintpot problem decades before.) And although many admirers of Bryan have framed his later anti-evolution crusade in light of his opposition to social Darwinism and eugenics,[62] it is important not to lose sight of the equally ardent opposition by many of Scopes's defenders and Bryan's opponents.

Bryan was a pious man with a fundamentalist understanding of the Bible and the world: "I would rather begin with God and reason down, than begin with a piece of dirt and reason up."[63] His faith alone would have likely been enough to make him an opponent of evolution, as had the beliefs of Samuel Wilberforce, Lord Kelvin (William Thomson), Louis Agassiz, and many of Darwin's other antagonists. But the crusade that he would soon launch, which we cover in the following two chapters, was also informed by Bryan's politics and morality, at a time when evolutionary science was popularly associated with social Darwinism and eugenics. And Bryan did nothing by half-measures; his campaign would be an effective one.

CHAPTER FIVE

Divergence

> I am simply trying to protect the word of God against the greatest atheist or agnostic in the United States.
>
> —Testimony of William Jennings Bryan, Trial of John Scopes, Day 7

BRYAN LOST HIS THIRD and final presidential campaign in 1908 to Roosevelt's successor (and soon-to-be opponent), William Howard Taft. By that time, Darrow stood as the country's preeminent labor lawyer, having honed his significant courtroom talents in defense of labor activists, including radicals accused of committing crimes of violence in support of their cause. His career trajectory shifted, however, when Darrow was himself accused of arranging to bribe a juror while defending men charged with a bombing during a Los Angeles labor action. In 1912, Darrow found himself in the unaccustomed role of defendant. Although not convicted, the charges frayed his relationship with the labor movement, especially after many former allies distanced themselves in his time of greatest need. From that point on, Darrow pivoted to lecturing, representing criminal defendants, and advocating on behalf of social issues dear to his heart. He still occasionally took on labor cases, but he became, first and foremost, a criminal defense attorney with a specialty in representing controversial defendants. He now traveled the road that would bring him to Dayton.

In contrast, Bryan's political career reached its zenith soon after Darrow was indicted. Long out of power, the Democrats received new life when Theodore Roosevelt ran a third-party challenge to Republican incumbent Taft. Darrow's old client, Debs, was by now also a perennial candidate for president; he received 6 percent of the vote, a high watermark for his socialist party. The crowded and chaotic ballot opened a narrow path for Democrat Woodrow Wilson, and Bryan maintained enough influence within his party to become Wilson's secretary of state.

It was an odd pairing from a personal perspective, even if the political logic added up. The urbane Wilson, whose uncle had been expelled from a South Carolina seminary in the 1880s after he endorsed Darwinism,[1] once observed that "whenever we discuss the structure or development of anything, whether in nature or in society, we consciously or unconsciously follow Mr. Darwin."[2] Later, he affirmed that of course he believed in evolution, "like every other man of intelligence and education."[3] The pious Bryan, on the other hand, refused to serve wine at state dinners, earning the derision of Washingtonians who scoffed at his "grape-juice diplomacy."[4]

But the Midwestern iconoclast now sat near the seat of power and found himself in a surreal conversation with the nation's elite. As secretary of state, Bryan crossed paths with entrepreneur, philanthropist, and Spencer-acolyte Andrew Carnegie.[5] And at a Washington dinner that included Holmes, the great justice discussed his prodigious reading list with the Great Commoner. Holmes told Bryan that, were St. Peter to ask him "have you ever read Gibbon's *Rome*," Holmes "would be very much embarrassed to admit" that he had not.[6] Considering that Holmes had rejected Christianity since at least the Civil

War, his lighthearted reference to St. Peter might have been a needle directed at the famously fundamentalist secretary of state. But Bryan does not seem to have taken it as such.

Bryan's encounter with power did not last long. When the First World War erupted in Europe in the summer of 1914, he staunchly opposed American intervention and unsuccessfully sought to mediate. Wilson initially sided with him but soon began to drift toward intervention. In response, Bryan resigned. Their split was not absolute—Bryan refused invitations to run against Wilson as a Prohibitionist in 1916. Indeed, that election gave progressives like Bryan new reasons to criticize the institution of the Supreme Court, after the ostensibly nonpartisan justice Charles Evan Hughes resigned to challenge Wilson on the Republican ticket. Bryan backed Wilson over Hughes and supported the Democratic line, which narrowly prevailed. Soon after Wilson's re-election, America entered the war. Although Bryan never abandoned his support for progressive reforms, from that point on, his political efforts focused on two cultural issues: prohibition and opposition to evolution.

As for Darrow, in contrast to many of his erstwhile allies in the radical labor movement, he saw Germany's invasion and harsh occupation of Belgium as emblematic of the strong subjugating the weak, implicating his core values. He supported the war wholeheartedly, although he would later come to regret the wartime crackdown on civil rights and would atone by representing political prisoners deprived of their civil liberties.

WORLD WAR

Civil liberties did not fare well during the war, or during the "first red scare," which followed the Russian Revolution of 1917. Congress enacted the sweeping Espionage Act,[7] which criminalized, in addition to traditional spying, the obstruction of military recruitment as well as the circulation of subversive materials. These provisions were interpreted broadly and used against many who spoke against the war, as well as members of radical movements. It was a time of overreach.[8]

Darrow initially supported the government by delivering speeches in favor of US intervention. He even met with Wilson in 1917. At that meeting, Darrow may have spoken up for his old allies in the radical labor movement, many of whose publications were being suppressed under the Espionage Act. Wilson promised in a letter to Darrow to "try to work out with the Postmaster General some course with regard to the circulation of the Socialistic papers that will be in conformity with the law and good sense."[9] It was a vague promise. Little changed.

Darrow's sometimes-friend and former client, Eugene Debs, was an obvious target. During the war, the nation's most prominent socialist was jailed for giving a pacifist speech alleged to have obstructed recruiting.[10] His conviction was upheld by the Supreme Court, as were those of other protestors. Justice Holmes, for one, showed little sympathy for Debs' pacifism; as a young man, Holmes had left Harvard to fight for the Union and been wounded three times in action against the Confederacy.[11] He penned the Court's harsh decision, now considered to be wrongly decided. From prison, Debs received nearly a million votes for president in the election of 1920.[12]

Darrow's support for the war had alienated his old client, and Debs refused an initial offer from Darrow to represent him. But Debs was by now old and in poor health, and after the war, Darrow came to regret his temporary de-emphasis of civil liberties. He diligently worked on the postwar campaign to secure his former client's early release. Debs wrote to Darrow soon after his sentence was commuted to time served by President Warren Harding, thanking Darrow for the "loving and inspiring words you uttered and the substantial services you rendered during my imprisonment, all of which will be gratefully remembered to the last of my days."[13] Although they had drifted apart over time and Darrow no longer primarily represented the radical labor movement, he did not hesitate to campaign for an old friend. And his recommitment to advocacy for individual liberty primed him for the trials that would come.

As Debs's post-war commutation suggests, normalcy had largely returned by the mid-1920s.[14] Indeed, wartime overreach arguably backfired, bringing some around to a more expansive view of civil liberties on the precipice of the *Scopes* trial. Holmes, for example, had largely held in favor of the government during the fighting. But he rebelled against government overreach shortly after the armistice, in *Abrams v. United States*, when the Court's majority upheld espionage convictions against Russian Americans who had distributed pamphlets critical of capitalism. In his dissent, joined by Brandeis, Holmes laid out an expansive vision of free speech that would, decades later, become mainstream: "the best test of truth is the power of the thought to get itself accepted in the competition of the market," and government accordingly should not prosecute speech unless it posed an imminent and immediate threat.[15] This famous maxim, which provided the roots of the now influential concept of the "marketplace of ideas,"[16] echoed Holmes's earlier and more explicitly Darwinian observation that the common law exemplified a "lively example of the struggle for life among competing ideas, and of the ultimate victory and survival of the strongest."[17] From a wartime crackdown came the seeds of an expansive view of speech rights—one that would be influenced by Holmes's longstanding interest in natural selection.[18] Expanded speech rights would, in turn, have direct relevance for legal outcomes related to the discussion of controversial ideas like evolution.

Darrow soon played a role in another seminal case with an impact on speech rights. Yet another controversial client was Benjamin Gitlow, a radical inspired by the Russian Revolution who was prosecuted by the state of New York for criminal anarchy. Gitlow insisted on addressing the jury directly, despite Darrow's advice, arguing before the court that capitalism was on the verge of falling to a socialist revolution. "I suppose a revolutionist must have his say in court, even if it kills him," Darrow conceded.[19] In closing, Darrow did his best, arguing that revolution was quintessentially American. He referenced the founders and John Brown. It was to no avail; Gitlow was quickly convicted.

At the Supreme Court, Gitlow's other attorneys argued that the freedom of speech enshrined by the *First Amendment* was also included within the *Fourteenth Amendment*'s reference to "liberty." The implications were enormous because the Fourteenth Amendment applied to state governments, whereas on its own, the First Amendment did not. In a decision released one month before *Scopes*, a majority of the justices considered this argument without rejecting it outright. Instead, they decided that Gitlow lost either way, holding that, even if the federal Constitution protected freedom of speech in a state criminal case, those speech rights offered no protection for Gitlow's leftist agitation.

Holmes dissented (joined, again, by Brandeis), reiterating his newfound stronger view of free speech and arguing that the "general principle of free speech, it seems to me, must be taken to be included in the Fourteenth Amendment, in view of the scope that has been given to the word 'liberty.'"[20] His dissenting view would become the law—his dissents often did—but not quite in time for *Scopes*. Eventually, the Fourteenth Amendment would be fully redeemed in the eyes of liberals. Although some progressives had argued for its repeal in the era of *Lochner*, it would come to be the central protector of civil rights against state governments during the era of the civil rights movement.

Apart from *Gitlow*, one other early-1920s case signaled a coming shift and laid the groundwork for litigation concerning cultural controversies and school curricula. In *Meyer v. Nebraska*,[21] the Supreme Court struck down state statutes that barred foreign-language instruction in public schools (enacted due to wartime hostility to all things German). Invoking the Fourteenth Amendment's concept of liberty, the Court held that it encompassed not just "the right of the individual to contract" but also more personal rights, such as the freedom to establish a home and bring up children. This time, speech rights were not mentioned, but the decision implied that, at least in the absence of a national emergency, the power of state legislatures to prohibit areas of educational study might be limited by the Constitution.[22]

CULTURE WAR

By this point, Darrow and Bryan were no longer allies or even fellow travelers. Bryan advocated for prohibition; Darrow was a vocal "wet." Bryan launched himself into a campaign against evolution; like many other liberals, Darrow forcefully criticized him for doing so. Between the two causes, most modern accounts of the final chapter of Bryan's life emphasize his devotion to anti-evolution causes more than his devotion to prohibition. Perhaps this is unsurprising—Bryan's dramatic death in Dayton inevitably connects the end of his life to his anti-evolution crusade. Moreover, efforts to remove evolution from schools continue unabated today, whereas efforts to impose temperance through the law have long been a dead letter. It is worth noting, however, that some contemporaries felt that the reverse was true. For example, the *St. Louis Post-Dispatch* wrote in 1923 that Bryan "has crusaded along many a tangent, but for a number of years prohibition has been his great objective."[23]

Bryan was, of course, not the progenitor of American religious campaigns against evolution; these found their roots in late nineteenth-century theological debates between liberal modernism and conservative fundamentalism, amplified by the widespread introduction of evolutionary theory in new areas of the country through the creation of universal, compulsory public school systems.[24] Several prominent evangelists, such as T. T. Martin, who campaigned as a leader of the "Anti-Evolution League of America," crisscrossed the country in response to the spread of the theory, seeking to "drive Evolution from all tax-supported schools."[25] But the movement that would merge a campaign for populist legislation with a defense of traditional religious doctrine and a challenge to perceived moral decay seemed tailor-made for Bryan, and he would devote much of the remainder of his life to it.

Despite his failures to achieve the presidency, Bryan was an effective advocate. He recognized and took advantage of the disarray in the evolutionary field, caused in part by the rise of mutationism and the temporary decline of Darwinian natural selection. For example, British biologist William Bateson, who had coined the term "genetics" in 1906 and helped to spread acceptance of Mendelism,[26] had long opposed Darwinian explanations of natural selection. At various points, Bateson made headlines by attacking Darwinism:

> Modern research lends not the smallest encouragement or sanction to the view that gradual evolution occurs by the transformation of masses of individuals . . . The great advances of science are made like those of evolution, not by imperceptible mass improvement, but by the sporadic birth of penetrative genius. The journeymen follow after him, widening and clearing up, as we are doing along the track that Mendel found.[27]

Gleeful that evolutionists were seeming to turn against Darwin, Bryan pounced. He quoted Bateson directly and presented his speeches as evidence of the general failings of evolutionary theory.[28]

Fundamentalist agitation against evolution was already spreading to public schools. In New Mexico, in 1922, a school superintendent named F. E. Dean was dismissed after a conflict with the local school board over evolution. Dean's former professor, University of Missouri zoologist Winterton Curtis, began a letter-writing campaign on Dean's behalf, reaching out to eminent individuals who had been mischaracterized as opponents of evolution by one of Dean's local rivals. Curtis wrote to Bateson, whom he had met, warning Bateson that his speeches had "been grossly misunderstood and misrepresented by the anti-evolutionists," and that "teachers in the high schools of the South and Southwest are being restricted in their presentations of the Evolutionary Doctrine and even dismissed from positions." Curtis solicited a response, writing that he would use it "both for Mr. Dean and to combat Mr. Bryan's propaganda." Bateson responded by reaffirming his belief in evolution despite his specific disagreement with Darwinian natural selection:

> We do know that the plants and animals, including most certainly man, have been evolved from other and very different forms of life. As to the nature of this process of evolution, we have many conjectures, but little positive knowledge.[29]

Curtis reached out to other prominent figures on Dean's behalf, including Woodrow Wilson—the president Bryan had briefly served under.[30] Wilson responded by letter in August 1922, writing, "of course like every other man of intelligence and education[. . . .] I do believe in organic evolution. It surprises me that at this late date such questions should be raised."[31]

Despite Curtis's campaign, Bryan continued to reference Bateson in anti-evolution speeches. In his planned closing argument for *Scopes*, for example, he included the statement that "Prof. Bateson came all the way from London to Canada to tell the American scientists that every effort to trace one species to another had failed."[32]

In some quarters, Bryan and his allies were gathering momentum. They found their first legislative success in Florida, a state Bryan had adopted (he became wealthy late in life in a Miami-based real estate boom). He had few allies left among northerners and liberals. In June of 1923, Bryan responded by letter to criticism of his Florida advocacy in the *Chicago Tribune*, emphasizing that he objected only to teaching evolution as a fact and arguing that his activism "recognized the injury that is done to students by the teaching of materialistic evolution." Bryan also defended his position on prohibition and accused the *Tribune* of "consistently ridicule[ing] the attitude of a majority of the Christian church."[33]

This was all too much for Darrow, especially appearing in his hometown newspaper. Darrow rebutted Bryan in the *Tribune*, asking Bryan an extended series of questions about literal interpretations of biblical stories that conflicted with modern science: Was the Earth made in six days? How did Noah gather animals from other continents? How could the sun have stood still for Joshua? And the like.[34] (The *Tribune* editors' own response jousted with Bryan over the miracle of Christ turning water into wine, needling him over his support for Prohibition.)[35]

For now, Bryan refused to engage with his former ally: "Mr. Darrow is one of two atheists with whom I am acquainted . . . The man who denies the existence of God is not likely to have much influence . . . atheism when avowed is not nearly so dangerous as so-called theistic evolution."[36] But soon, Darrow would be putting virtually the same questions to Bryan as the former secretary of state took on the witness stand. It would be the last battle of Bryan's life.

Between Bryan and Darrow's clash in the pages of the *Chicago Tribune* and their rematch in small-town Tennessee, Darrow took on one of his most controversial engagements yet, a high bar for a lawyer who had represented activist dynamiters and communist agitators. In 1924, Darrow vigorously defended the country's most notorious murderers, two teenagers from wealthy Chicago families by the names of Leopold and Loeb. The pair had coldly and brutally slain a young boy from a neighbor's family in a profoundly disturbing attempt to demonstrate their supposed ability to commit the perfect crime.[37] In accepting the representation, Darrow was motivated by his fierce opposition to capital punishment. Execution seemed the teenagers' certain fate.

In arguing against the death sentence, Darrow deftly crafted an argument rooted in biological, psychological, and philosophical theories of human behavior, contending that the teenagers' actions had been predetermined by their nature. He decried the legal system for assuming "human conduct was not influenced and controlled by natural laws," charging that the system ignored scientific causes of crime. Although quintessentially Darrow, this argument echoed a passage by Holmes, who had once reflected on the possible legal implications of scientific theories suggesting behavior was determined by natural factors.[38] "We lawyers go on and on and on," Darrow decried, "punishing and hanging and thinking that by general terror we can stamp out crime." For Leopold and Loeb, he won life sentences, successfully sparring them from execution.[39]

In modern times, Darrow's courtroom performance has been described as brilliant and held up as a gold standard for attorneys defending clients facing execution.[40] It had contemporary admirers as well. Senator La Follette (whose court-reform plan had been a culmination of progressive criticism of the judiciary) asked Darrow for a written copy of his plea and wrote that he would "cherish . . . to read [it] again and again."[41] But praise was far from universal. Many focused on the horrific nature of the crime and saw the result as tragic. As far away as Kentucky, prosecutors opined before a jury that "all a man has to do at this day and time is to hire the best lawyers to be had, and you will remember the miscarriage of justice that we read about a few months ago . . . in the case of Loeb and Leopold, two of the foulest murders in recent history, and justice is defeated."[42]

Perhaps unsurprisingly, Bryan numbered among the critics, and he would soon make Darrow's nature versus nurture arguments against the death penalty a centerpiece of his own attacks on evolutionary theory. In a speech published shortly after the *Scopes* trial (based on his planned closing argument), he wrote:

> Psychologists who build upon the evolutionary hypothesis teach that man is nothing but a bundle of characteristics inherited from brute ancestors. That is the philosophy which Mr. Darrow applied in this celebrated criminal case. "Some remote ancestors"—he does not know how remote—"sent down the seed that corrupted him." You cannot punish the ancestor—he is not only dead but, according to the evolutionists, he was a brute and may have lived a million years ago . . . Evolutionists say that back in the twilight of life a beast, name and nature unknown, planted a murderous seed and that the impulse that originated in that seed throbs forever in the blood of the brute's descendants, inspiring killings innumerable, for which the murderers are not responsible because coerced by a fate fixed by the laws of heredity![43]

Darrow's argument was pioneering in introducing psychological and biological factors into legal defense, touching upon determinism and the complex interface between biological understandings of behavior, such as aggression, and their legal and ethical implications. Many of these questions remain relevant and challenging today: how are behaviors that might be explained through and influenced by evolutionary processes interpreted and managed within a legal framework, and how do they relate to questions of diminished capacity? Bryan's response was a perceptive and effective rejoinder to the outer limits of arguments rooted in biological determinism. Of course, lost in Bryan's rhetoric is that Darrow was arguing only against the application of the death penalty, not against holding Leopold and Loeb criminally responsible for their actions.

Meanwhile, Bryan's ongoing campaign for anti-evolution laws continued unabated. In the beginning of 1923, he visited Dartmouth College to speak against evolution. This visit was largely shaped by the efforts of Professor William Patten, who introduced evolutionary theory into the curriculum as early as 1898 through his course on Comparative Anatomy of Vertebrates. By 1921, Patten made evolution a required course for all freshmen, an unprecedented move arising from Patten's belief that evolution was a unifying concept that linked not only scientific fields but all realms of human thought. However, the introduction of this course was met with significant resistance from alumni, faculty, students, and the wider public, many of whom saw it as a challenge to religious values. Patten persisted, with strong support from Dartmouth's president, Ernest Hopkins, and over time, the opposition waned.[44]

At Dartmouth, Bryan delivered a lecture titled "Science vs. Evolution." His staunch defense of biblical creationism and his criticisms of evolutionary theory were well-received by some, but many Dartmouth students remained unmoved by his arguments, distinguishing between his oratorical talent and the logical shortcomings of his position. Attending students were asked, before and after the event, to rank their acceptance of evolution on a scale of one (rejecting the doctrine completely) to five (accepting it completely). Of the 136 attendees, only two rejected it completely, and seventy accepted it completely. After hearing the Great Commoner speak, the number rejecting evolution to some extent increased from seven to fourteen, but most student comments were negative: "A skillful, masterly oration, given by a prejudiced man, who did not have . . . a knowledge of what he was talking about."[45]

Bryan had, by this time, become a member of the American Association for the Advancement of Science (AAAS). At that year's meeting of the association, the former president of the body's zoological section, Edward Loranus Rice, devoted his speech to attacking Bryan. Rice, much like Asa Gray before him, was a devout scientist, and he especially criticized Bryan's position that evolution was irreconcilable with faith, urging greater accommodation between the two fields. Rice praised Darwin's meticulous collection of evidence—although in his own work, he had rejected Darwin's theory of sexual selection (the explanation for the peacock's tail, among other extravagant traits), as "relegated by many to the rank of a somewhat doubtful hypothesis rather than theory."[46]

As he had done with Bateson, Bryan attacked whenever he saw disagreement within the evolutionary field (perhaps explaining why he had joined the AAAS in the first place). In a later speech, for example, he charged that "sexual selection has been laughed out of the class room, and natural selection is being abandoned."[47] And on the fifth day of the trial, he would label sexual selection as one of the "absurdities of Darwin," claiming that it had been discarded.

Bryan would soon find another legislative success, outraging Darrow as much as Darrow's arguments for Leopold and Loeb had outraged Bryan. These were different men then they had been in 1896, when their paths had brought them to the Democratic convention during a surge of enthusiasm for silver and anger at the treatment of Eugene Debs after the Pullman strike. Their focuses had narrowed, and one issue that captivated both of their attention—evolution—both would soon consume the nation's as well.

CHAPTER SIX

A Magnificent Opportunity to Test an Obnoxious Law

> I do not know how about where these foreign gentlemen come from, but I say this in defense of the state . . . the most ignorant man of Tennessee is a highly educated, polished gentleman compared to the most ignorant man in some of our northern states, because of the fact that the ignorant man of Tennessee is a man without an opportunity, but . . . the northern man in some of our larger northern cities have the opportunity without the brain.
>
> —Argument of Sue Hicks, Trial of John Scopes, Day 5

A TENNESSEE REPRESENTATIVE NAMED J. W. BUTLER helped give Bryan his next and most famous legislative victory. Butler represented a rural area of Tennessee, about sixty-five miles from Nashville. "We are poor folks in this section of the country," he would tell an interviewer. "But we believe in God and we believe in the Bible."[1]

The fundamentalist backlash against evolution was gaining steam, urged on not only by Bryan but also by anti-evolutionist campaigners like T. T. Martin and the popular evangelist (and former Chicago Cubs outfielder) Billy Sunday. "Old Darwin is in Hell," pronounced Sunday, who left the majors to pursue ministry.[2]

Influenced by this growing religious response, Butler introduced what would become Tennessee's anti-evolution law, soon to be known as the Butler Act. Local concerns about education reform influenced state politics as well,[3] but this law was primarily part of a national backlash against evolution—exemplified by Bryan's previous victory in Florida, the fight over Curtis's former student in New Mexico, and several states that would follow Tennessee's example after *Scopes*. The Butler Act prohibited any teacher in any public university or school "to teach any theory that denies the story of the Divine creation of man as taught in the Bible, and to teach instead that man has descended from a lower order of animals." Despite the advice of Bryan, who opposed criminal penalties, violation of the Butler Act was made a misdemeanor, with a fine ranging from one hundred to five hundred dollars.[4]

There was, by this point, a yawning divide between regions like Dayton and the nation's more liberal and cosmopolitan quarters with respect to the acceptance of evolution. The extent to which even non-scientists at elite institutions had embraced Darwinian theory is exemplified by a remarkable 1919 *Yale Law Journal* essay by Yale professor A. G. Keller, titled "Law in Evolution."[5] Keller argued that, much like species in the natural world, human laws and institutions evolve in response to changing environmental conditions and pressures: that cultural "mores are as much evolutionary products as are the horse's hoof and the camel's foot."[6] His argument (which anticipated sociobiological debates of the late twentieth century) reflected some of the limitations and biases of early twentieth-century thought, particularly in its uncritical engagement with ideas derived from Social Darwinism. It also demonstrated the cultural chasm that had come to divide even non-scientific disciplines like law, with prominent scholars deeply engaged in attempts to apply evolutionary theory even while Southern legislators and rural judges would view it with suspicion.

Not surprisingly, the reaction to the Butler Act at elite institutions was fierce. The dean of Columbia's College of Pharmacy urged all universities to refuse to recognize educational credentials from any state that had enacted an evolution ban.[7] Meanwhile, a relatively young legal advocacy group, the American Civil Liberties Union (ACLU), saw in anti-evolution laws a potential test case—an opportunity to win a favorable judicial decision striking down the law and setting a precedent that would advance the organization's liberal agenda. The ACLU began seeking a client and promising a "friendly test case" that would not cost a teacher his job. One such ad found its way to Rhea County.

DARWINISM IN DAYTON

Rhea County, Tennessee, defied easy stereotypes, although many would be pressed upon it by the reporters who descended on Dayton in 1925. (H. L. Mencken, for one, admitted in print that he had "expected to find a squalid Southern village, with . . . pigs rooting under the houses and the inhabitants full of hookworm and malaria."[8]) The county had been transformed by railroads in the 1870s when track was laid through its lands to connect Chattanooga to Cincinnati. The resulting line ran four miles west of the county seat of Washington, threatening the town's relevance and continued viability. The county's response reflected the power and importance of railroads at the time—it simply moved the county seat. A hamlet of two hundred lay along the new route. The county courthouse in Washington was promptly torn down and rebuilt there, in Dayton, Tennessee.[9]

Dayton briefly flourished when British investors organized the Dayton Coal & Iron Company (whose company scrip practices were taken to the Supreme Court) to manufacture pig iron and coke. Despite its name, the firm was not exactly *of* the town. Its managing director was based in Glasgow, Scotland, and was a member of the firm's British parent company. Its highest-ranking American officer worked in Cincinnati, and its legal counsel resided in Chattanooga.[10] Nevertheless, boosted by this foreign investment, Dayton enjoyed a minor boom around the turn of the century. The year 1906 saw the construction of the town's first public high school.[11] Dayton also stood apart as a rare Republican stronghold in the deep South. Ironically (given the local support he would enjoy at the trial), as a Democratic presidential candidate, Bryan carried Tennessee but lost Rhea County each time he ran.[12]

But the modest boom times did not last especially long. Dayton Coal & Iron's British parent failed in 1913, leading to its insolvency and dragging down the town's economy. The firm was still being wound down in 1925, with its remaining property managed by a court-appointed receiver.[13] After over ten years of stagnation, local businessmen were searching for a way to put their once-promising town back on the map.

One of these men was George Rappleyea, an entrepreneurial chemist at a rival coal and iron firm, also in decline. A northerner by birth and a scientist by profession, Rappleyea had no sympathy for the fundamentalist movement's effort to enforce traditional religion through legislation. When the ACLU publicized its plans to challenge the Butler Act, Rappleyea saw an opportunity to bring attention to Dayton. For it to work, he needed to present the ACLU with a willing defendant, and he found the right man for the job.

Those who knew him described John Scopes as unassuming and humble. A native of Kentucky, his family had relocated to Salem, Illinois, when he was a teenager. Salem had one significant claim to fame: it was the hometown of William Jennings Bryan, who had been raised on a nearby farm (his father, an attorney, having believed a farm to be the ideal place to raise a family).[14] In 1919, a few years after he departed Wilson's cabinet, Bryan returned to Salem to deliver the commencement address at Scopes's high school graduation. Scopes and another classmate had laughed at a shared joke during the speech, briefly distracting the legendary orator. He and Bryan both remembered the incident, sharing this point of connection when they met—amicably—during the trial six years later.[15]

After his memorable graduation, Scopes came to Tennessee to study at university, later taking a job as a football coach, general science teacher, and occasional substitute teacher in Dayton. At Rappleyea's suggestion, he volunteered to be arrested for violating the Butler Act, although he could not have predicted the storm that was coming.[16] Rappleyea convened a meeting between Scopes, himself, and local officials at Robinson's Drug Store, a local meeting point. There, a "trial of the century" was set into motion.

Far from Dayton, in the ACLU's New York City headquarters, the key lawyer organizing the effort was Arthur Garfield Hays. A principled liberal, by his own account Hays had an "inclination to sympathize with dissenters."[17] Hays was utterly fearless; years after the *Scopes* trial, he would travel to Germany during the rise of the Nazis, despite risks arising from his Jewish heritage, to help defend a man accused of involvement with the Reichstag fire.[18] When Rappleyea contacted the organization, Hays saw "a magnificent opportunity to test an obnoxious law."[19]

Whether the ACLU approved or not, Scopes himself played a role in choosing his own defenders. Scopes quickly agreed to also be represented by a local attorney: John Randolph Neal, a law professor and former Tennessee politician. Neal was a Rhea County native and the son of a Confederate officer. Yet he was also an intellectual and a supporter of modern public education. As a Tennessee state legislator, Neal had authored bills providing generous funding for the state's public schools and libraries.[20]

Neal's interest in the case went beyond his belief in public education. Two years earlier, he had been dismissed from his position as a law professor at the University of Tennessee after coming to the defense of a psychology professor who in turn had been dismissed after teaching evolution while the Butler Act was being debated in the legislature. Sources differ about whether either dismissal was directly caused by evolution controversies; Neal's employment, in particular, was also jeopardized by his ongoing friction with the dean, highly unorthodox self-presentation, and eccentric refusal to cash his university paychecks.[21] But for Neal, the *Scopes* prosecution echoed his own dismissal and doubtlessly had a personal element.

As publicity grew around the coming test of the Butler Act, Bryan volunteered his services for the prosecution, announcing a rhetorical "duel to the death" against evolution. He brought along his son, William Jennings Bryan Jr., a federal prosecutor in California who agreed to assist his father. Although the Bryan father-and-son team would take center stage, the prosecution would be nominally led by Tom Stewart, a young Tennessee attorney general and a lawyer of some talent. (Stewart went on to become a US Senator in the 1930s.) With Stewart were Tennessee attorneys Gordon McKenzie and Sue Hicks,[22] neither of whom did much to distinguish themselves in the courtroom.[23]

With Bryan joining the prosecution, his former ally turned rival was almost certain to volunteer for the defense. Hays was pleased; he and Darrow were friends and worked well together. However, his colleagues at the ACLU had a cooler reaction. Those who hoped for a careful test case narrowly focused on constitutional debate foresaw that, with Darrow and Bryan involved, the trial was sure to become a maximalist clash between science and religion.

ACLU legal director Roger Nash Baldwin had a skeptical eye toward religion. Born to an old New England family of "agnostic Unitarians," he had discovered books by Darwin and Huxley as a young man and developed an interest in evolution. Darwin's account of the *Beagle*'s voyage was a favorite.[24] Nevertheless, Baldwin balked at Darrow's inclusion; he adamantly opposed allowing the *Scopes* trial to become a replay of the Huxley-Wilberforce debate—or an extension of Darrow's heated newspaper arguments with Bryan. Baldwin hoped instead to emphasize a constitutional right to free speech and academic freedom as the central issue of the case. Another ACLU board member, Walter Nelles, sought to frame the case around the separation of church and state. Both wanted a less controversial face for the defense team.

As one alternative, Baldwin advocated for Charles Evan Hughes, the former Supreme Court justice who had resigned to challenge Wilson, by then working as a practicing attorney.[25] Hughes shared Baldwin's view that anti-evolution laws were problematic because they infringed on the "freedom of learning." As a former Supreme Court justice himself, his presence would have certainly aided the ACLU in their ultimate goal of reaching the high court and winning there. Had he become involved instead of Darrow, the *Scopes* proceedings and their legacy might have been entirely different. The spectacle of two unsuccessful presidential candidates squaring off in Dayton might have been a different sort of curiosity, but Hughes was not the unorthodox firebrand that Darrow was. In later remarks on *Scopes*, Hughes explicitly avoided the subject of evolution in order to speak "in a non-controversial spirit."[26] Needless to say, avoiding controversy was not Clarence Darrow's way.

Hays was close to both Darrow and Baldwin (he later dedicated a book to both men)[27] and he stood up for Darrow when

Baldwin and others opposed his inclusion. After a spirited discussion, Hays got his way within the ACLU, and Darrow joined the team. Baldwin, perhaps annoyed, stayed away from the proceedings his organization had helped set in motion.[28]

Also joining the *Scopes* defense team was New York attorney Dudley Field Malone, who had once served in Wilson's State Department under Bryan. Like his former boss, now turned opponent, Malone had broken with Wilson. He left federal service and opposed US involvement in World War I, believing that America should stay out of the bloody conflict consuming Europe. Malone endorsed a Socialist party candidate after the United States entered the war, advocating for a non-punitive peace with Germany. At a speech during the war, his audience had accused him of treason, although, unlike Debs, he had avoided any Espionage Act prosecutions.[29] After a failed campaign for the New York governorship, Malone was now practicing as a successful divorce lawyer. He agreed to come to Dayton and work with Darrow.

This remarkable lineup of out-of-state attorneys began to arrive in town, and Dayton started to receive the attention that Rappleyea had wanted. According to Hays, the town "took on the character of a revivalist-circus. Thither swarmed ballyhoo artists, hot dog venders, lemonade merchants, preachers, professional atheists, college students, Greenwich village radicals, out-of-work coal miners, I.W.W.'s, Single Taxers, 'libertarians,' revivalists of all shades and sects, hinterland 'soothsayers,' Holy Rollers, an army of newspaper men, scientists, editors and lawyers."[30] Even advertisers got in on the publicity: "There is no 'Monkey Business' connected with the evolution of STUDEBAKER Automobiles," proclaimed a car ad in the *Knoxville Sentinel*.[31] J. W. Butler, the sponsor of the law that had created this controversy in the first place, grumbled, "I can see no reason for bringing all these celebrated lawyers and others in to argue a case that any Tennessee lawyer could handle as well as any of them."

While the ideological battle lines were becoming clear, the legal arguments against the Butler Act, and the role that evolutionary science would play in the courtroom, were not. The initial internal dispute within the ACLU had involved more than just skepticism over Darrow's outsized personality. The organization got the test case it had wanted, but it had yet to precisely articulate which principle was going to be tested. The law's challengers had two competing, serious theories of the case—and long after the case was over, reasonable minds would disagree which was stronger.

To be sure, the defense advanced *many* arguments about the Butler Act's invalidity in their initial motions to dismiss the indictment. Several turned on specific provisions of the Tennessee Constitution. For example, the defense contended that the Butler Act violated a rule governing the relation between a law's caption and its substance.[32] Likewise, the state constitution directed Tennessee's government to "cherish literature and science," and Neal would argue that "in no possible way can science be taught or science be studied without bringing in the doctrine of evolution."[33]

But the ACLU had not spent its time and money engineering a test of the Butler Act just to win a decision based on particular provisions of the Tennessee Constitution. Any such ruling, if favorable, would apply only within Tennessee. Moreover, the Tennessee Supreme Court retained the final say when interpreting and constructing the state's own constitution. To the extent that the ACLU's sights were set on the US Supreme Court—and the possibility of winning a victory with significance for the entire country—the defense would ultimately need to argue that the Butler Act infringed on an individual right that the *federal* constitution protected against interference by state governments.

The Fourteenth Amendment—specifically, its provision guaranteeing that life, liberty, or property may not be infringed by state governments without due process of law—was the best candidate. State laws such as New York's maximum working-hours regulation (in *Lochner*) and Nebraska's prohibition against teaching German (in *Meyer*) had been struck down as interferences with liberty that were so substantively unjustified that the relevant state legislatures had denied due process to their constituents. The Butler Act might be attacked in the same way, but the defense would need to define and articulate a liberty that it infringed before attempting to place that liberty within the scope of the Fourteenth Amendment. Two contenders—core liberties specifically articulated by both state and federal constitutions—rose above the rest: the freedom of speech and the freedom of religion.

ACADEMIC FREEDOM

ACLU director Roger Nash Baldwin believed that the primary defect of the Butler Act was its interference with academic freedom. The argument assumed that a teacher's classroom speech enjoyed constitutional protection against interference from governmental authorities. Presumably, the liberty of a teacher to speak freely in a classroom would be a subset of the broader freedom of speech, specifically enshrined in the federal Constitution's First Amendment and in analogous provisions in each state constitution. As Neal would put it during the defense's initial motions: the Butler Act infringed on the "freedom of expression of a man's ideas and a man's thoughts."[34]

There were seeming advantages to this view of the case. For one thing, it might allow Scopes's defenders to stake out a less controversial position, avoiding polarizing questions about the tension between Darwin's works and a literal interpretation of the Bible. Indeed, a freedom-of-speech argument could conceivably avoid any engagement at all with the ongoing disagreements about how evolution occurred: The freedom to speak does not turn on whether that speech is scientifically validated.

Moreover, a speech-based theory seemed compatible with the Supreme Court's recent decision in *Meyer v. Nebraska*—perhaps the best controlling precedent for the defense—in which a seven-justice majority reversed the conviction of a teacher for violating a wartime law prohibiting German-language instruction. *Meyer* had not expressly mentioned speech generally or academic freedom specifically, but it had outlined several liberties protected by the Fourteenth Amendment, including "the right of the individual to contract, to engage in any of the common occupations of life, to acquire useful knowledge, to marry, establish a home and bring up children, to worship God according to the dictates of his own conscience, and generally to enjoy those privileges long recognized at common law as essential to the orderly pursuit of happiness by free men."[35] Although the decision is susceptible to multiple readings, the majority in *Meyer* seemed to take the view that Nebraska's law had violated the *teacher*'s right to *teach* (to engage in "the common occupations of life"). If Meyer had a right to teach German, perhaps Scopes had a right to teach evolution.

However, this theory of *Scopes* as a *free speech* case also faced significant hurdles. A major barrier was another Supreme Court decision (issued just over a month before Scopes went to trial) against Darrow's former client, revolutionary leftist Benjamin Gitlow. There, the Court had a chance to announce the very rule that Baldwin was seeking: a holding that the freedom of speech, as articulated in the First Amendment, is one of the liberties protected by the Fourteenth Amendment against interference by state legislatures. In dissent, Holmes and Brandeis adopted precisely this position. But the other seven justices dodged the issue, holding that Gitlow's radical manifestos were unprotected regardless of whether the Fourteenth Amendment applied to the case. Thus, as of 1925, only two justices had been willing to definitively rule that the federal right to freedom of speech applied to the states as well as the federal government. And two out of nine was not a winning coalition.

And it is far from clear that this theory of the *Scopes* case would even be persuasive to both of the two dissenters in *Gitlow*. Although Holmes had voted in Gitlow's favor, he had also been one of the dissenters in *Meyer*, and Holmes was particularly highly unlikely to be sympathetic to an argument framed around academic freedom for public school teachers. Although Holmes personally placed far more faith in science than religion, he had a long history of deferring to popular legislation even when it contradicted his personal preferences. "[I]f my fellow citizens want to go to Hell I will help them," he once wrote, in a personal letter. "It's my job."[36]

Moreover, Holmes's narrow view of the rights of government employees—a class that, of course, included public school teachers—was influenced by the legal maxim that the "greater includes the lesser." This interpretive principle suggested that whenever an actor *expressly* has the power and discretion to take an action, that actor *impliedly* has power to condition that action or act in a lesser way. For example, the explicit power of a governor to pardon a convicted criminal includes an implied power to commute the offender's sentence to time served; the explicit power of a city to pave new streets includes an implied power to resurface existing ones; and so on.[37]

This style of legal reasoning is very old,[38] and uncontroversial examples abound. But it was refined and deployed by Holmes to such an extent that modern scholars and judges often refer to it as "Holmesian reasoning."[39] The eminent justice lent his influence to greater-includes-the-lesser arguments in many decisions dating back to his time on the Massachusetts Supreme Judicial Court before he was elevated to the US Supreme Court.[40]

Holmesian greater-includes-the-lesser reasoning eventually became controversial due to its negative implications for individual rights. Specifically (and with great relevance to *Scopes*), it implied that, because no one is *entitled* to public employment, the government can thus condition its workers' employment on their agreement to refrain from exercising their rights. In other words, the greater power to deny a government job altogether includes the lesser power to place conditions on continued employment. As usual, Holmes put it simply and memorably: A policeman "may have a constitutional right to talk politics, but he has no constitutional right to be a policeman."[41] Could the same be said for a public school teacher talking about evolution?

There was yet one more important flaw with Baldwin's vision of *Scopes* as a speech case: if a goal of defense was to protect and advance the teaching of evolution, then a holding based on the right of a teacher to free expression within a classroom would be, at best, a double-edged sword. Such a right would protect a teacher who explained evolution notwithstanding a state directive to refrain from doing so—but it would also protect a teacher who wished to buck an official curriculum to decry evolution. With the benefit of hindsight, we now know that many cases involving evolution in public schools arise in the latter posture, with an anti-evolutionist teacher determined to teach creationism despite a school district's curricular decision to teach evolution. Governmental control over education can work in two directions.

In sum, free speech arguments would play a significant role in the *Scopes* defense, but the theory was tactically and strategically flawed. A competing line of argument held more promise.

THE SEPARATION OF CHURCH AND STATE

The other strong argument that the Butler Act violated the federal Constitution claimed that it contravened protections for religious liberty. The First Amendment, and most state constitutions, guarantee the right to freely exercise religion while prohibiting the government from establishing religion—the latter mandate is frequently referred to as the separation of church and state. As of 1925, no decision had yet incorporated these religious rights into the Fourteenth Amendment. But in *Meyer*, the Court had referenced the freedom "to worship God according to the dictates of [a person's] own conscience" as one of the liberties protected by the First Amendment.[42] Arguably, that could be interpreted as a reference to the totality of the First Amendment's clauses directed toward religion.

To be sure, the Butler Act did not directly establish a state religion or prohibit religious worship. But the defense would argue that the law prohibited the teaching of a *scientific fact* merely because of an objection rooted in particular religious doctrines. Characteristically, Darrow argued that point in the most intense and maximalist way:

> If today you can take a thing like evolution and make it a crime to teach it in the public school . . . [s]oon you may set Catholic against Protestant and Protestant against Protestant, and try to foist your own religion upon the minds of men. If you can do one you can do the other. Ignorance and fanaticism is ever busy and needs feeding. . . . After a while, your Honor, it is the setting of man against man and creed against creed, until with flying banners and beating drums we are marching backward to the glorious ages of the sixteenth century, when bigots lighted fagots to burn the men who dared to bring any intelligence and enlightenment and culture to the human mind.[43]

This theory of the case, however, clashed against the state's general power to control the curriculum taught in its public schools. As Stewart would argue, "[y]ou can attend the public schools of this state and go to any church you please," but "this is the authority, on the part of the legislature of the state of Tennessee, to direct the expenditure of the school funds of the state."[44] At least within reason, Stewart had a point. A state can uncontroversially select which textbooks to adopt or which books to purchase for a school library. It can add or subtract particular topics from required curricula, and it can determine that certain theories merit general instruction while others do not. The Constitution does not prohibit a state government from deciding to place, for example, more emphasis on calculus than social studies, or vice versa. Yet to many, the state crosses a line when these decisions are made with religious motivation. These arguments are still being made today, and not just with respect to evolution but also with respect to other subject matters that generate cultural controversy, such as education on sexual and gender identity.[45]

More broadly, defense arguments against the power of the state to determine the subjects in its curriculum cut against the grain of progressive lawyering at the time, a notable development given the left-wing orientation of many of Scopes's attorneys. Movement lawyers had spent decades arguing that courts should be *more* deferential to the decisions of state legislatures. They had fiercely criticized cases like *Lochner*, where judges had evoked the Fourteenth Amendment to restrict the power of elected representatives and invalidate their attempts to improve the lives of their constituents.

This tension was not lost on the liberal press: As the *New Republic* fretted during the trial: "The power now exercised by American courts and particularly by the Supreme Court to prohibit legislative experiments for the real or alleged reason that they violate some utterly ambiguous higher law, will in the long run provoke more labor and anxiety for American progressives than all the anti-evolution statutes that are likely to be passed."[46] If the courts should have shown more deference to the legislature in *Lochner*, why should they show less in *Scopes*? And if Holmes's famous dictum—that the Fourteenth

Amendment did not enact the works of Spencer—was indeed an apt response to the *Lochner* majority, could it also be said that the Fourteenth Amendment did not enact the work of Darwin?

The answer to these questions and problems lay where the science of evolutionary biology would intersect most directly with law. The defense would not (and could not) argue that a legislature lacked any power to control its schools' curriculum. Rather, it was curricular modification *with a religious purpose* that was improper. And to distinguish the Butler Act from other valid and reasonable exercises of state power, the defense would seek to establish that the exclusion of an established scientific fact from a science curriculum was arbitrary and irrational—and thus evidence that the law was enacted for a constitutionally invalid religious agenda. Although a peripheral or uncertain subject could presumably be dropped or excluded for any number of rational reasons, evolution was (according to the defense) a necessary component of modern scientific understanding and thus could not be rationally excluded.

Malone would thus argue:

> There is no branch of science which can be taught today without teaching the theory of evolution, and that this applies to geology, biology, botany, astronomy, medicine, chemistry, bacteriology, embryology, zoology, sanitation, forestry and agriculture.[47]

It was a bold claim; almost certainly hyperbolic with respect to, for example, astronomy. But Judge Raulston understood the argument, acknowledging that whether the Butler Act "was a reasonable exercise" of the state's power "would depend largely on whether evolution is a mere guess by a few men, or generally accepted by all scientists."[48] Raulston's statement implied that the case might depend on the validation of evolutionary science in the courtroom.

This was precisely what the defense wanted. The defense team planned to present expert testimony from a litany of scientific experts. Technical scientific subject matter might not have been the entertainment that Rappleyea and other boosters had envisioned—a correspondent covering the trial for the *Manchester Guardian* observed that Dayton's "townspeople are little interested in the discussion of scientific questions as compared with religious" ones.[49] But here the defense's strategy was consistent with the progressive legal movement's broader embrace of judicial education. By bringing a vast team of scientific experts to Dayton, the defense was attempting a version of what Brandeis had done with his famous brief in *Muller*: compiling discoveries from outside the law to place the narrow dispute between the parties in a broader context. Whether or not they would succeed, as Brandeis had, remained to be seen.

A NARROW FOCUS

For his part, Tom Stewart preferred to leave grand discussions of evolution and religion to the media. His theory of the case was as simple as theories can be: Scopes had broken a law. Objections that the law was unwise, reactionary, or out of line with modern science should be addressed to elected legislators, not a local jury or a circuit judge. He thus appealed to the same intuitions about the democratic process that animated progressive objections to *Lochner* and its prodigy.

Accordingly, Stewart sought to strip down the proceedings, moving to dismiss all extraneous arguments. For most of the trial, he succeeded, convincing Judge Raulston to keep most of the defense's legal argument, expert testimony, and scientific stipulations out of the ears of the jury. As for the ACLU, it had been planning an appeal from the beginning, and its lawyers were likely largely indifferent as to whether expert testimony was heard by the Dayton jury or simply included in the record.

Stewart's plan to keep the trial narrowly focused would almost succeed. But in the end, it would be fatally undermined by his scene-stealing co-counselor. It had, in fact, been many years since William Jennings Bryan genuinely practiced law. Nevertheless, Bryan's stature and rhetorical talents were simply too great for him to be marginalized by the likes of Stewart, McKenzie, or Hicks. He had come to Dayton for his "duel to the death" with evolution (and perhaps, in a more personal way, with Darrow). He would get it. But before the trial's legendary culmination, the defense's pivotal contentions about the indispensability of evolution would be put to the test.

CHAPTER SEVEN

Evolution in the Courtroom

> The general principle of evolution has nothing to do with natural selection. The later might be totally discredited without in the least shaking the validity of the principle.
>
> —Argument of Arthur Garfield Hays, Trial of John Scopes, Day 5

NO ONE REALLY EXPECTED Judge Raulston to quash the indictment of Scopes by ruling in favor of the defense's pre-trial motions. Had it occurred, that result might have annoyed the ACLU most of all, given that the small town trial had been designed to create a favorable binding precedent on appeal, not to win immediate exoneration of a defendant who had agreed to be indicted.

Raulston was no Oliver Wendell Holmes, but he analyzed the defense's claims in a Holmesian manner. Because the state of Tennessee was not compelled to hire anyone as a school teacher, it was within its rights to hire teachers on conditions, such as the condition that an employee not teach evolution. This was greater-includes-the-lesser reasoning on display. As Raulston announced:

> There is no law in the State of Tennessee that undertakes to compel this defendant or any other citizen to accept

> employment in the public schools. The relations between the teacher and his employer are purely contractual and if his conscience constrained him to teach the evolution theory he can find opportunities elsewhere.[1]

The arguments the defense made in its initial motions would be preserved for appeal, but for now, the trial would continue.

But before evidence would be presented, the trial opened (in an ominous sign for the defense) with a prayer. Throughout the coming days, Darrow would repeatedly and strenuously object to the judge's practice of opening proceedings with official prayers, to no avail. A jury was then selected, and that process did not bode well for Scopes. During preliminary questioning of the panel, Darrow asked one potential juror, a pastor, whether he ever preached for or against evolution. "Why, I preached against it, of course!" the pastor replied, drawing applause from the audience in the courtroom.[2]

THE SCIENTISTS

The many experts who had traveled from the nation's great universities to testify were perhaps the most out-of-place group in Dayton—a high bar given the circus-like atmosphere surrounding the trial. The illustrious group included zoologist Maynard Metcalf, a professor at Johns Hopkins; Kirtley Mather, chairman of the geology department at Harvard; Fay Cooper-Cole and Horatio Newman, an anthropologist and a zoologist, respectively, from the University of Chicago; and Winterton Curtis, a zoologist from the University of Missouri. For the latter, at least, his involvement had a personal element; it had been Curtis who involved himself in a proto-*Scopes* debate in 1922 when a former student had been discharged for teaching evolution in New Mexico. At the time, Curtis had solicited and received a letter in support of evolution from William Bateson and Woodrow Wilson.[3]

The defense had trouble finding sufficient housing in Dayton. Ultimately, they rented and worked out of an old mansion overlooking town that they dubbed a "haunted house." It became a kind of dormitory, housing the lawyers and expert witnesses. A police squad was brought in from Chattanooga to provide security for the visiting scientists in case the group was targeted by the Ku Klux Klan. During the trial, an intense storm came through Dayton as the witnesses and defense attorneys conferenced in the haunted house. "Boys, if lightning strikes this house tonight . . ." Darrow joked.

But although they had amassed a superb team to defend the field of evolution, in 1925, the field itself was still laboring through the confusion created by declining acceptance of natural selection. That confusion permeated the defense's explanations of evolution. Even had the defense argued in a friendlier forum—and before a judge who allowed them to speak to the jury—their presentation of evolution as an unquestioned scientific consensus was deeply undermined by the state of the field, as they described it.

As arguments began, Malone engaged in an extensive analogy between evolution and embryology, arguing that the evolution of a species proceeded in a way that mirrored the cellular development of an individual organism: "Evolution never stops from the beginning of the one cell until the human being returns in death to lifeless dust."[4] Here, Malone was guided by a theory of German zoologist and Darwin devotee, Ernst Haeckel (1834–1919), who had posited that "ontogeny recapitulates phylogeny." Haeckel's "biogenetic law," formalized in 1866, held that the developmental stages of an organism (ontogeny) replay the evolutionary history of its species (phylogeny). For example, Haeckel suggested that during the development of a human embryo, the organism passes through stages that resemble forms of its evolutionary ancestors, such as fish, amphibians, and reptiles. He believed this mirrored the sequence of species evolution. Malone gave an example: "at the end of one month, the [human] embryo . . . has gill slits." The implication was that embryological development provided a window into the evolutionary past.[5] The theory would be referenced at the trial several more times (although it later came to be regarded as oversimplified and wrong in critical ways).

The initial witness for the defense was zoologist Maynard Metcalf, a professor at Johns Hopkins. Darrow began by asking him about his understanding of evolution. In his answer, Metcalf again conflated the evolution of species and the embryological development of an organism: "I have always been particularly interested in the evolution of the individual organism from the egg, and also of the evolution of organisms as a whole from the beginning of life."[6]

Darrow then asked whether Metcalf would say that all men of science were evolutionists. Metcalf answered that, from his personal knowledge, virtually all zoologists, botanists, and geologists who had made substantial contributions to their field believed in evolution. However, he also made a critical qualification: "I doubt very much if any two of them agree as to the exact method by which evolution has been brought about, but . . . I know there is not a single one among them who has the least doubt of the fact of evolution." Metcalf would return to this theme later in his testimony:

> There are dozens of theories of evolution, some of which are almost wholly absurd, some of which are surely largely mistaken, some of which are perhaps almost wholly true, but there are many points—theoretical points as to the methods by which evolution has been brought about—that we are not yet in possession of scientific knowledge to answer. . . . I think it would be entirely impossible for any normal human being who was conversant with the

> phenomena to have even for a moment the least doubt even for the fact of evolution, but he might have tremendous doubt as to the truth of any hypothesis.[7]

From a legal perspective, Metcalf's testimony undermined the defense. Could it really be beyond the police power of a state to remove a subject from a high school curriculum, if even the top experts in the field could reach no consensus regarding the mechanisms through which that subject operated?

After Metcalf, the prosecution counterattacked. Advancing his narrow conception of the case, Stewart moved to exclude the testimony by these scientists, arguing that their viewpoints were irrelevant to the only question at issue: whether Scopes had broken the law as it stood on the books. He asserted that "they cannot come in here and try to prove that what is the law is not the law."[8]

The prosecution's objection to the admissibility of scientific evidence was skillfully argued by William Jennings Bryan Jr. The younger Bryan deftly summarized many of the problems posed by the admission of any expert witness testimony in a courtroom. He explained that although experts are "absolutely necessary" in cases where complex technical issues must be explained to juries, they nonetheless create a danger that the jury may "substitute [the expert's] opinion for their own." In his view, because the fact that Scopes had taught evolution had been conceded by the defense, there could be no technical issues of fact that expert opinion would illuminate. Instead, they created an unjustified danger of improperly influencing the jury's ultimate view of facts that had already been established.[9]

Here, Bryan Jr. tapped into an ongoing controversy over the appropriate role of expert witnesses that had, ironically, been intensified by Darrow's brilliant and controversial performance the year before at the trial of Leopold and Loeb.[10] John Wigmore, a titanic scholar of evidence (who, in keeping with the times, had adopted an "explicitly evolutionary theory of jurisprudence"[11]), responded to the trial of Leopold and Loeb with an essay warning that, rather than serving "the cause of Science and Justice," the current, lax system created "scandal and mistrust that now attaches so often to expert testimony."[12] Wigmore's essay was published just months before *Scopes*.

Hays countered by arguing that expert witnesses were indeed required to understand the application of the Butler Act to the facts of the case, and his response says much about the state of evolution at the time:

> The title of the act refers to evolution in the schools . . . The secret difficulty lies in the fact that there are two Darwinisms, the popular one and the technic[al] one. The laymen uses Darwinism as a synonym of evolution in the broadest sense; the evolutionist never uses the word in this sense, but always uses it as a synonym for natural selection, one of Darwin's chief theories. The general principle of evolution has nothing to do with natural selection. The later might be totally discredited without in the least shaking the validity of the principle.[13]

Hays was right that Darwin's name had come to have two meanings: both as a general stand-in for evolution and as a specific way of referring to Darwin's proposed mechanisms for evolution. And this was a problem for evolutionists at the time—Darwin was inseparably associated with the field, but many of his specific ideas were out of fashion. Again, the defense had partially conceded the rather chaotic state of evolutionary theory at the time.

It was during these pivotal arguments on the admissibility of scientific evidence that the elder William Jennings Bryan made his first contribution to the case. Bryan's arguments, along with those of his fellow prosecutors, won the day, and Judge Raulston ruled the defense's expert witness testimony inadmissible. Yet he allowed them to be entered into the record, in case the appellate court disagreed with his evidentiary ruling.[14] Thus, although the jury was not ultimately exposed to the scientific testimony offered by the defense, the judge, attorneys, and the appellate court were—along with the many Americans who studied the transcript, which was rushed to publication (and sold under the title *The World's Most Famous Court Trial*) after the sensational case.

The scientific evidence submitted in the record represented a diverse range of fields and offered a broad array of positions supporting evolution. Kirtley Mather, chairman of the geology department at Harvard, outlined archeological finds that provided evidence for the evolution of humans. These included a fragment of a jawbone showing teeth similar to those of great apes and humans, dating from between two and ten million years ago; the skull of an organism, dubbed the "Ape Man of Java," with features nearer to that of a human, dating from between one and two million years before; and fossils of Cro-Magnons (early modern humans), dated to forty or fifty thousand years before. Each of these bore progressively greater resemblance to modern humanity. Such findings pointed to incremental evolution over the course of millions of years. Mather asserted, "Such facts . . . can be explained only by the conclusion that man has been forced through long processes of progressive development, . . . traced backward through successively simpler forms of life."[15]

Anthropologist Fay-Cooper Cole of the University of Chicago explained the similarities between anthropoid apes and humans, demonstrating how they must share a common ancestor. He asserted that it is "impossible to explain anthropology or the pre-history of man without teaching evolution."[16] Wilbur Nelson, the state geologist of Tennessee, discussed geological finds in the state, explaining that fossils of simpler organisms were found in more ancient rocks, with progressively more complex fossils found in rocks from later eras.[17] This, of course, contradicted the idea that all life was created simultaneously.

Zoologist Horatio Newman of the University of Chicago pointed to the vestigial structures in the human body, such as the remnants of a tail, gill slits in the human embryo, and functionless muscles analogous to those that move the ears in other species. These "and numerous other structures can be reasonably interpreted as evidence that man has descended from ancestors in which these organs were functional." He also discussed recent discoveries of the fossils of early descendants of man—including "Piltdown Man," once thought to establish the "missing link" between apes and humanity, but later discovered to be a forgery.[18]

Zoologist Winterton Curtis of the University of Missouri, not engaged in his first or last[19] battle in defense of evolution being taught in schools, focused on the debate between evolutionary biologists concerning the mechanisms of evolution and the means of natural selection. He conceded that many contemporary scientists questioned Darwin's theory of natural selection, but he cautioned that this did not mean they questioned the validity of evolution as an overarching concept. Curtis asserted that the evidence for the latter was overwhelming.[20] Given his long-term commitment to the defense of evolution, Curtis's role at trial was ultimately less than he had hoped; the next month, a former student wrote to him that "I am sorry for your sake that things quieted down so soon at Dayton."

The prosecution's own witnesses made less of an impression. Stewart first called the school's superintendent to introduce the textbook that Scopes taught from, George William Hunter's *A Civic Biology: Presented in Problems.*[21] Hunter's *Civic Biology*, first published in the previous decade, was a major textbook used around the country.[22]

The book ever so briefly explained Darwin's theory of natural selection and featured an evolutionary tree that would play a significant role in the case. Neither linear nor hierarchical, the textbook's tree depicted classes of animals as circles, each circle's size drawn to indicate the number of species in the corresponding class. Each circle was connected, like a balloon, back to the original class of protozoa, the "strings" branching off from each other at points of evolutionary divergence.[23]

To the outrage of the anti-evolutionists, humanity was not shown separately; rather, there was only a (relatively small) circle for the class of mammals. Nor did mammals stand alone at the "top" of the tree—they were roughly aligned with birds. In what was intended to be his closing argument, instead delivered as a speech shortly before his death, these features drew the ire of Bryan:

> *No circle is reserved for man alone*. He is, according to the diagram, shut up in the little circle entitled "Mammals," with thirty-four hundred and ninety-nine other species of mammals. Does it not seem a little unfair not to distinguish between man and lower forms of life? What shall we say of the intelligence, not to say religion, of those who are so particular to distinguish between fishes and reptiles and birds, but put a man with an immortal soul in the same circle with the wolf, the hyena and the skunk?[24]

This was the crux of the anti-evolutionist reaction to Darwinian views—and it remains so today. As the textbook accurately depicted, mammals, including humanity, share more traits with each other than with other classes of animals such as reptiles or birds because common ancestors of today's mammals diverged from the ancestors of today's reptiles and birds long before mammalian species differentiated from each other.

But although the textbook's evolutionary tree conveyed these important points, the text also contained many serious and harmful errors. It thus discussed "five races or varieties of man" as if these were distinct categories from an evolutionary standpoint, and listed Caucasians as "the highest type of all." These were not views universally held by contemporary evolutionary biologists; as already noted, Asa Gray, to name one, had pressed the opposite conclusion. And scientific racism had also been deployed by intellectual *opponents* of evolution, such as Jenkin. Nevertheless, its association with evolution in such a widely used textbook likely led many to draw the connection.

As already noted, *Civic Biology* also contained content on eugenics, including error-filled case studies supposedly demonstrating the hereditary nature of negative traits within families. These sections had, in fact, drawn complaints to the publisher's Boston office in 1915; these complaints largely focused on the explicit (by the standards of the time) discussions of human reproduction rather than the broader social and moral implications of eugenics.[25] To be sure, there was no evidence that Scopes taught from these sections of the textbook, nor were they directly referenced in the trial. But their presence in Hunter's textbook serve as a stark reminder of competing and controversial implications of the developing understanding of evolution.

After introducing the textbook and engaging in a cursory examination of a handful of witnesses, including students from Scopes's class, the prosecution had introduced all the evidence it needed to establish that Scopes taught evolution. So long as the Butler Act was valid, there was no doubt that Scopes had broken it. And Raulston had already rejected the motions challenging the Act's constitutionality. The trial could have ended there, and if it had, it would almost certainly have been forgotten.

But the trial would produce one last surprise, which we cover in the next chapter. Afterward, it would proceed through a strange appeal, falling short—for an unexpected reason—of the ACLU's goal of reaching the Supreme Court.

CHAPTER EIGHT

Conviction

> We see nothing to be gained by prolonging the life of this bizarre case.
>
> —*Scopes v. Tennessee*, 289 S.W. 363 (Tenn. 1927)

THE CULMINATION OF THE TRIAL—the clash between Darrow as attorney and Bryan as an expert on the Bible, fought over the age of the Earth, the Tower of Babel, Heliocentrism, and Cain's wife, all over the vehement but ineffective objections of Stewart—need not be exhaustively recounted here.[1] The exchange was broadcast. It was breathlessly reported in the world's newspapers, from the *Baltimore Evening Sun* to the *North China Daily News*. It quickly became an American legend and has been reenacted on the stage and on screens. But although their examination became the most famous part of *Scopes*, it was also more general than the previous arguments. It was untethered to the debates within evolutionary biology or to the legal implications of scientific consensus. Moreover, Darrow and Bryan essentially rehashed in-person debates they previously held over the pages of newspapers. This did much to entertain but little to enlighten.

Darrow's focus was on biblical passages where a literal interpretation of text stretches credulity—could the sun have "stood still," for example, as it is said to have done in the Book of Joshua? Bryan waivered a bit; he conceded, for example, that the days of creation in Genesis might refer to "periods" instead of twenty-four-hour days. Repeatedly, he simply disclaimed interest in the scientific implications of his beliefs. Although he largely stood his ground, his defense was deemed weak by most observers. H. L. Mencken derided what he considered to be Bryan's "palpable imbecilities."[2]

After their fierce witness stand theatrics, the defense rested without closing argument. The clever maneuver deprived Bryan of the chance to redeem himself by delivering his own closing, which he had carefully prepared and planned as his final, fiery rebuttal to Darrow's case and Darwin's theory. With the case prematurely submitted to the jury, Bryan was left with no choice but to deliver his extensive final argument as a post-trial speech to the public. Strategically, the defense lost nothing: besides the court of public opinion, their true goal was always a win on appeal, where closing argument to the jury would be largely irrelevant. As anticipated, Scopes was quickly convicted, and Raulston sentenced him to the minimum fine under the statute: $100.

The national media departed; few stuck around for Bryan's post-trial speeches. The Great Commoner, however, did not rush away from Dayton. At this point, he was in poor health, beset by poorly managed diabetes. He arranged for the publication of his post-trial speech as a written argument—it would serve as his reply to the growing media narrative claiming he had been made to look foolish by Darrow. As it happened, it would also be his final address to the nation. Bryan succumbed to his health problems on July 26, 1925, just five days after he took the stand. Virtually all national journalists had left Dayton so quickly after the trial that they missed the opportunity to report on his death.

LEGACY

Modern historians have partially rehabilitated the image of Bryan at the trial.[3] But most depictions of his contributions at Dayton, particularly in the aftermath of *Scopes*, tended to be unflattering or worse. The hugely influential political historian Richard Hofstadter, for example, wrote that Bryan "had long outlived his time" and displayed a "childish conception of religion" and "inchoate notions of democracy."[4]

To be sure, Bryan had repeatedly deployed explicitly and evocatively Christian rhetoric, and by doing so, perhaps undercut the state's argument that its actions did not afford an impermissible preference to any particular religion. Bryan did little to reassure on that front when he argued that "this principle of evolution disputes the miracle . . . and they believe man has been rising all the time, that man never fell."[5] But he was hardly alone in defending the Butler Act through explicitly religious argument. His co-counsel, McKenzie, had argued at trial, "[W]e didn't have any sun until the fourth day [of creation]. I believe the Biblical account."[6] Stewart also dismissed appeals to religious pluralism, and a question about how the Butler Act related to Muslims, by disclaiming that this was not "a nation of heathens." Thus, although Hofstadter and others depicted Bryan as an avatar of religious fanaticism, his express appeals to the dogma of the Christian faith differed little from those of the other prosecutors.[7]

But alone among his fellow prosecutors, Bryan made an attempt to engage with evolutionary theory and even appeared to intuitively recognize some of the inconsistencies in the field as it stood in 1925. Far from the buffoonish figure described by Mencken and Hofstadter, he responded to the scientific arguments presented by the defense more perceptively and subtly than did Stewart, McKenzie, or Hicks.

For one, Bryan cleverly zeroed in on the defense's repeated conflation of the development of an individual organism with the evolution of species—and he arguably explained the difference between the two with more precision than did Scopes's lawyers. By doing so, he illuminated the way that the defense's case had inadvertently revealed the uncertain status of the evolutionary consensus in the 1920s. Although the defense's attorneys and experts had good reason to analogize between embryology and evolution,[8] Bryan backed into a cogent critique in his description of student Howard Morgan's testimony about Scopes's lesson:

> [Darrow] thought that little boy was talking about the individuals coming up from one cell. That wouldn't be evolution—that is growth, and one trouble about evolution is that it has been used in so many different ways that people are confused about it. . . . No wonder [Darrow] was not able to distinguish by just hearing it once, between the evolution of life that began in the ocean away down in the bottom and evolved up through animals bigger and bigger . . . That is evolution and that is what he taught. Not the growth of an individual from one cell, but the growth of all life from one cell.[9]

Bryan's adeptness was also demonstrated in a colloquy with Darrow when the two debated the concept of selflessness and reciprocity. Bryan seemed to intuitively understand that the concept of natural selection posed a question: why would a willingness to sacrifice one's self for others unrelated by blood—found in all human societies and in many social animals—be an adaptive trait selected for through evolution? Bryan pointed to the teachings of Christ to help people "in proportion to their needs" rather than in proportion to the amount they might help one's self. He contrasted Christianity favorably with other doctrines of altruism, which he characterized as being rooted in reciprocity. "Reciprocity is a calculating selfishness," Bryan stated.[10]

Perhaps he had answered his own question. Strikingly, in this characterization, Bryan foreshadowed a later expansion of evolutionary theory that would be developed in the 1970s, when Robert Trivers formalized the idea that altruism among unrelated individuals could indeed be explained by reciprocity.[11] Bryan intuitively grasped that altruism could be explained through "selfish" reciprocity (and thus be an adaptive trait), although he sought to distinguish this form of altruism from Christian teachings.

Most significantly, Bryan was the only prosecutor to cogently address the moral implications of some interpretations of Darwinism. Cleverly (albeit inaccurately), he did so in part by reference to Friedrich Nietzsche—and to Darrow's arguments the preceding year, during the trial of Leopold and Loeb. In his brilliant and impassioned argument against the death penalty, Darrow had famously attributed the murder to the impressionable Leopold's understanding of "Nietzsche's philosophy of the superman" and his interpretation that "the intelligent man was beyond good and evil." Darrow had not, of course, blamed Nietzsche or those who taught his works, but he had presented Leopold's reaction to the teaching as one of many mitigating facts.[12]

Nietzsche was a natural foil for Bryan; the German philosopher's famous assertion that "God is dead"[13] had become one of the most iconic and extensively debated declarations in philosophical discourse. Bryan labored to link Nietzsche to Darwin, arguing that Nietzsche had listed Darwin as one of the "three great men" of the nineteenth century, and—implicitly invoking Spencer as well—asserting that Nietzsche's "supermen were merely the logical outgrowth of the survival of the fittest with will and power, the only natural, logical outcome of evolution."[14] Then, speaking (inaccurately) of the First World War but unintentionally presaging the Second, Bryan asserted that "Nietzsche gave Germany the doctrine of Darwin's efficient animal in the voice of his superman, and . . . the military textbooks in due time gave Germany the doctrine of the superman translated into the national policy of the super-state aiming at world power."[15]

However, the specifics here were not quite right. Nietzsche was indeed influenced by Darwin but rejected many of his ideas and specifically criticized the theory of natural selection.[16] Again, the phrase "survival of the fittest" and its application to human society originated with Spencer, whose conception of evolution was largely influenced by Lamarck rather than Darwin. These malign concepts (especially German militarism!) were hardly natural outgrowths of Darwin's ideas. Nevertheless, Bryan cleverly tied Darrow's past assertions about the power that a misinterpreted philosophy might have on an impressionable mind to Bryan's own concerns about social Darwinism, eugenics, and public morality.

In his intended closing argument, Bryan also criticized eugenics, seeking to connect it to evolution. He decried that its supporters' "only program for man is scientific breeding, a system under which a few supposedly superior intellects, self-appointed, would direct the mating and the movements of the mass of mankind—an impossible system!"[17] Again, in some ways, this was unfair and irrelevant. Scopes had certainly not advocated for eugenics, and his attorneys each had impeccable backgrounds as champions of individual liberty. And yet, Bryan's critiques cannot be dismissed out of hand given the history of how evolutionary concepts were deployed to justify oppressive policies throughout the early twentieth century. His arguments at *Scopes* were not always scientifically accurate, but he skillfully pointed to issues that the scientific community had yet to fully resolve, and he refused to allow moral considerations to be brushed aside. Without him, the trial's legacy would not have been nearly as durable.

APPEAL

The ACLU now had its chance to bring its case to a higher court. Unfortunately, the appeal was quickly bungled. The defense sought to contest Raulston's exclusion of scientific testimony "for the purpose of explaining the theory of evolution, the facts upon which that theory is based and the scientific accuracy and authority therefor."[18] Yet the defense failed to certify and file the record on time: Raulston had given them thirty days to do so at the close of trial, but Scopes's attorneys appear to have been under the impression that they had a sixty-day window instead.[19]

This degree of incompetence from a grouping of elite (and in some cases, legendary) lawyers is difficult to explain. It may have been largely Neal's fault; as local counsel, filing the record would have likely been his responsibility, and his career reveals an eccentric and rather erratic personality. Even so, it remains odd that the other attorneys, who had invested so much time and effort into this litigation, were not sufficiently involved to catch the error.

Upon their late filing, the prosecution quickly, and successfully, moved for portions of the record—including much of the excluded scientific witness testimony—to be struck. The team had failed, again, to formally present its evidence in support of evolution and was left in the unenviable position of asking the court to use its discretion to take judicial notice of facts not included in the technical record.

On appeal, the defense raised each of the arguments it had pressed in its trial motion. It also attempted to add a few more (a practice usually disallowed by appellate courts). For example, the appeal contended that the Butler Act violated a federal constitutional prohibition against states impairing contractual obligations, pointing to contracts between the federal government and the state that were premised "upon the condition that the teaching of the sciences should be promoted." However, the majority of the defense's appellate brief focused on the core theories of academic freedom and the separation of church and state. As for the latter, the defense's appellate brief included statements by J. W. Butler and Tennessee's governor explicitly referencing their religious motivations for enacting the law.[20]

The defense's appellate brief is perhaps most interesting for its extensive arguments that the Butler Act violated the Fourteenth Amendment. Citing cases including *Gitlow* and *Meyer*, the defense pushed back against the greater-includes-the-lesser argument[21] that "Scopes was not obliged to teach in the public schools," and "if he did not like the laws" affecting the public schools, "he might teach in private schools." The defense pointed to the Fourteenth Amendment's guarantee that "liberty" not be deprived without due process of law. "While the Supreme Court of the United States has refrained from defining the term 'liberty,'" the brief wrote, the rights of "teacher, pupil, and parents" were all violated by "arbitrary" legislation.[22] The brief contended that the Butler Act was arbitrary for a variety of reasons, including that it violated "public policy as expressed in provisions for religious liberty,"[23] a proto-incorporation argument, similar to the modern doctrine that the Fourteenth Amendment "incorporates" most provisions of the Bill of Rights and applies them to state as well as federal law.

On the other hand, the defense's appellate brief also fell into the trap inherent with a free-speech view of the case. The brief suggested that the statute discriminated by declaring "that candid evolutionists are guilty of crime if they state in the schools the doctrine they believe, while fundamentalists can proclaim their opinions without fear of punishment."[24] But of course, if *Scopes* were to prevail on this theory, the resulting precedent would seem to give fundamentalist teachers an equal and opposite freedom to ignore a school curriculum that *included* evolution.

The tenor of the state's opposition brief was harsh and biting. Stewart remained in charge, though William Jennings Bryan Jr. remained involved in the case (perhaps taking the remaining proceedings more personally after his father's death). Legally, the prosecution attacked the defense's incorporation arguments: "There is no provision of the Federal Constitution relating in terms to religious freedom that has any application to state legislation."[25] But tonally, its opposition brief was overwrought, filled with cultural signifiers and attacks on the opposing counsel rather than their ideas. It referred to Scopes's defenders as "Glib and garrulous," "charmed . . . by the flow of their own discourse,"[26] and set in the belief "that our legislature represents a very backward people not possessing *their* very cultured, intellectual and 'scientific' views."[27] Later, it charged that the law was under attack from "the discordant masses of anarchists, left-wing socialists, atheists, agnostics, self-styled 'intellectuals,' and 'scholarly Christians,'" noting that many who attacked the law "reside beyond the limits of Tennessee."

The Tennessee Supreme Court proved unsympathetic to the defense. A decision by Chief Justice Grafton Greene, issued in January 1927, asserted that "Evolution, like prohibition, is a broad term. In recent bickering, however, evolution has been understood to mean the theory which holds that man has developed from some pre-existing lower type." Applying this definition to the statute, the court found the Butler Act to be clear and unobjectionable.

The court then rejected the defense's arguments rooted in Scopes's constitutional rights, applying a version of the greater-includes-the-lesser maxim, stating that Scopes

> was under contract with the state to work in an institution of the state. He had no right or privilege to serve the state except upon such terms as the state prescribed. His

> liberty, his privilege, his immunity to teach and proclaim the theory of evolution, elsewhere than in the service of the state, was in no wise touched by this law. The statute before us . . . is an act of the state as a corporation, a proprietor, an employer. It is a declaration of a master as to the character of work the master's servant shall, or rather shall not, perform.[28]

The final nail in the coffin of the ACLU's case, however, came from the court's ultimate disposition of the case. Addressing a point that neither side had raised, the court held that Raulston had erred in setting the fine (rather than allowing it to be set by the jury). This was an odd and artificial reason to overturn this conviction—Raulston had given Scopes the minimum fine allowed by the statute, so there was no ground to complain that a jury had not weighed in to potentially set a higher fine. But the court took advantage of a pretext to end the case with finality, affirming the statute's validity while leaving Scopes uninjured and thus unable to seek any relief from the US Supreme Court. The ACLU's grand test case was unceremoniously stopped in its tracks.

Only one justice of the Tennessee Supreme Court seemed to take much interest in the battle between science and religion that had brought Darrow and Bryan to the state. Justice Alexander Chambliss had been born during the Civil War; his father was a Baptist minister serving as a chaplain in Robert E. Lee's Confederate army.[29] In a concurring opinion, Chambliss distinguished between "theistic" evolution and "materialistic" evolution. The former he saw as "held to by numerous outstanding scientists of the world," whereas the latter "seeks in shadowy uncertainties for the origin of life." Chambliss complemented the work of Scopes's lawyers but asserted that the Butler Act was nonetheless valid because, in his view, it barred only the teaching of "materialistic" evolution: "In this view the constitutionality of the act is sustained, but the way is left open for such teaching of the pertinent sciences as is approved by the progressive God recognizing leaders of thought and life." This concurring opinion was the best that Scopes's team would get from the legal system.

By the time the decision was issued, Darrow had moved on. In 1926, he won his last major case. Darrow successfully defended Ossian Sweet, a Black doctor who was charged with murder after he defended his family from a mob enraged because the Sweet family had moved into a previously all-white Detroit neighborhood. As for John Scopes, he declined various movie deals and lecture engagements in the immediate aftermath of his blockbuster trial. Instead, he departed Tennessee to study geology at the University of Chicago. Later, he would travel and work in South America, periodically sending updates about his experiences to Darrow, whom he had come to consider a friend.[30] His fame lasted only a short time, and he received little attention between the 1930s and the 1960s.[31]

The journey to understand evolution, however, was far from over. In the next chapter, we outline the process whereby the various unresolved questions that weighed on the defense's presentation at trial were addressed in the following decades. As a result of these developments, when the nation turned its attention back to evolution in education during the Cold War, it found a much more unified field.

A young Charles Darwin, whose theory of evolution through natural selection sparked enduring debates.

Wikipedia Commons

Exterior of Robinson's Drug Store in Dayton, Tennessee.

Courtesy of the Rhea County Historical Society,
Dr. Richard Cornelius Collection

Dayton's Main Street in 1925.

Courtesy of the Rhea County Historical Society, Dr. Richard Cornelius Collection

George Rappleyea, the engineer who helped initiate the Scopes Trial.

Courtesy of the Rhea County Historical Society, Dr. Richard Cornelius Collection

Clarence Darrow and William Jennings Bryan, opposing legal giants.

Courtesy of the Rhea County Historical Society, Dr. Richard Cornelius Collection

William Jennings Bryan with Judge John T. Raulston, presiding over the Scopes Trial.

Courtesy of the Rhea County Historical Society, Dr. Richard Cornelius Collection

William Jennings Bryan advocating for biblical creationism during the trial.

Courtesy of the Rhea County Historical Society, Dr. Richard Cornelius Collection

John T. Scopes (center) with defense attorneys Clarence Darrow (right) and Dudley Field Malone.

Courtesy of the Rhea County Historical Society, Dr. Richard Cornelius Collection

The seven scientific experts prepared to testify as to the validity of evolution. Back row, left to right: Horatio Hackett Newman, Maynard Mayo Metcalf, Fay-Cooper Cole, Jacob Goodale Lipman; Front row, left to right: Winterton Conway Curtis, Wilbur A. Nelson, William Marion Goldsmith.

Smithsonian Institution Archives

Winterton Conway Curtis, who contributed to science education before, during, and after the trial.
Smithsonian Institution Archives

Attorney John R. Neal with John T. Scopes.

Courtesy of the Rhea County Historical Society, Dr. Richard Cornelius Collection

Tom Stewart, prosecuting attorney.

Courtesy of the Rhea County Historical Society, Dr. Richard Cornelius Collection

F.E. Robinson, owner of the drug store where discussions leading to the Scopes Trial took place.

Courtesy of the Rhea County Historical Society, Dr. Richard Cornelius Collection

An anti-evolution billboard in Atlanta, Georgia in 2024, highlighting ongoing resistance to evolutionary theory.

Authors' photo.

CHAPTER NINE

Synthesis, Resurrection, and the Shadow of *Scopes*

> One trouble about evolution is that it has been used in so many different ways that people are confused about it . . . there were so many different kinds of evolution or so many definitions of evolution that . . . it would be useless.
>
> —Argument of William Jennings Bryan, Trial of John Scopes, Day 5

As the scientific testimony presented at *Scopes* made clear, Darwinism—in particular, its key concept of natural selection—was in steep decline by 1900 and was disavowed by several experts during the *Scopes* trial. Mutationism was ascendant as an evolutionary creative force, despite a growing body of research on organisms (such as fruit flies) indicating that most mutations harm individual vigor and survival.

This would all change very shortly after the *Scopes* case was decided by the Tennessee Supreme Court. Just a few years later, new ideas emerged that led to the downfall of mutationism. The field of evolutionary biology underwent a profound transformation during those years, when previously diverse or even opposing fields, including paleontology, population biology, Darwinian natural selection, and the fast-developing science of genetics were integrated to provide a comprehensive framework for understanding evolutionary processes. Had *Scopes* taken place just a decade later, the scientific rejoinder to anti-evolution controversies would have been vastly different.

The gains in theoretical understanding during this period, now labeled the "Modern Synthesis,"[1] were substantial, but these new conceptualizations became widely accepted only gradually, in part because of the mathematical complexity of the newly consolidated approaches.[2] Awareness of the vast relevance of the Modern Synthesis for connecting the many and diverse existing biological subfields[3] did not come automatically either but developed over time.

The genesis and consolidation of the Modern Synthesis was driven by a small group of central figures and their key contributions, greatly impacting post-*Scopes* debates over evolution.[4] Not surprisingly, some disagreements among the principal players appeared and persisted for years, but an eventual consensus nonetheless emerged on most of the foundational points by the mid-1950s—in time for another major wave of political and legal conflict over evolution that would begin in the late 1960s. When the battle would be joined again, in courtrooms in the latter half of the century, the evolutionary field was far more unified and confident.

DARWIN'S SAVIORS AND THE BEGINNING OF THE EVOLUTIONARY MODERN SYNTHESIS

Ronald A. Fisher (1890–1962) was, by 1930, Britain's leading statistician and had revolutionized that field with an earlier book.[5] In that year—just five years after the *Scopes* trial—he launched the Modern Synthesis by publishing *The Genetical Theory of Natural Selection* (1930),[6] dedicated to Charles Darwin's son, Leonard, one of Fisher's significant mentors.[7] In this transformational work, Fisher demonstrated that Darwinian natural selection, combined with Mendelian inheritance, is entirely consistent with the evolutionary vision Darwin had presented in *Origin*. Additionally, Fisher extended the range of biological phenomena that could be explained with these principles. His formal approach faced initial resistance due to its mathematical complexity, but his interpretations of his findings' relevance to evolution, and human evolution in particular, were much clearer. Fisher focused on the impact of small changes in large populations and consequently has been described as "the most thorough-going and uncompromisingly strict Darwinian among early synthesists."[8]

Fisher's book restated Darwin's fundamental theory of natural selection in a mathematical form, ignoring the contaminating and flawed theory of blending inheritance (the "paint-pot problem") that had plagued Darwin for years. Fisher's population genetics models provided a clearer understanding of how advantageous traits and the genes that underpin them could spread and become more prevalent over generations. These models showed how, when a favorable mutation appears, its change in frequency in successive generations would depend on the population size, the number of generations considered, the mutation rate, and the advantage conferred on bearers (among other factors). The core of evolutionary change could be modeled as a change in gene frequencies.

Fisher also formulated the "Fundamental Theorem of Natural Selection," one of the most widely cited ideas in evolutionary biology. In simple terms, it states that the rate at which a population's fitness increases is directly proportional to the genetic variation related to fitness in that population. Fitness, loosely, is the ability of an individual to survive and reproduce in its current environment. More formally, evolutionary biology is concerned with "relative fitness." Relative fitness measures an individual's reproductive success—the genetic contribution to the next generation—compared with other individuals (or other alleles) in the same population. It is not an absolute measure of how well an organism is adapted to its environment in isolation; rather, it is a measure of how well an organism's traits enable it to outcompete and reproduce more successfully than others in its population. Fisher's theorem provided a mathematical foundation for understanding the role of genetic variation in evolution, highlighting the importance of genetic diversity as the substrate for adaptation and underscoring the role of natural selection in shaping the characteristics of populations over time.

These ideas also provided new life to the concept of sexual selection. This idea had never gained much traction up to that point; Bryan said it had been "laughed out of the class room,"[9] and many scientists had indeed been skeptical. But Fisher outlined how extravagant traits, like the peacock's tail, could evolve. A trait (such as a male's long tail) might initially add to fitness (for example, by conferring advantages in terms of, for example, flight maneuverability). But over time, a female preference for long tails could come about as a result of random mutation. At that point, both the female preference and the fitness advantage will select for long tails. At a later point, tail length will cross a threshold where it no longer confers a flight advantage, but will be selected for purely as a matter of female preference. Fisher's mathematical models demonstrated this possibility.[10]

Around the same time, major contributions were also made by John Burdon Sanderson Haldane (1892–1964), a versatile and formidable scholar.[11] Haldane's work broadly addressed questions of how natural selection effects change on the genetic constitution of a population. He argued that "a satisfactory theory of natural selection must be quantitative"[12] (a position, as we have seen, that likely would have made Darwin squirm). Haldane's seminal contributions to the Synthesis included a book, *The Causes of Evolution* (1932), in which he argued that mutation "cannot prevail against natural selection of even moderate intensity,"[13] a claim he developed formally in a mathematical appendix. The book was less challenging than Fisher's for general readers, and it was the first to extensively survey a wide array of biological topics and integrate Mendelian genetics into an evolutionary framework.

Haldane made significant contributions to an understanding of the importance of genetic recombination.[14] He developed mathematical models to describe how the frequency of recombination between linked genes depends on the physical distance (i.e., the "linkage") between them. These models laid the foundation for the development of genetic mapping techniques, which identify the locations of genes on a chromosome and the distances between them.[15]

Haldane also introduced the concept of "genetic load," which considers the cost imposed on populations by the presence of deleterious mutations. The cost here is a reduction in overall fitness: on average, individuals in the population are less fit (lower reproductive success, increased mortality, or a decreased ability to compete for resources) because of these harmful mutations. Haldane mathematically modeled the impact of these mutations on the fitness of individuals within a population. Essentially, Haldane's measure of genetic load equals the relative chance that an average individual will die due to its disadvantageous genes before reproducing.

Later in his career, Haldane addressed the challenging issue of explaining how new advantageous mutations could spread throughout a population if the process of natural selection depends on rare favorable mutations. His calculations revealed what would later be termed "Haldane's Dilemma," that the rate of substitution of advantageous alleles was likely much slower than what had been imagined, thus posing a significant challenge to the rate of adaptive change.[16] Haldane is also remembered as one of the first theorists to recognize the importance of kinship (specifically the genes shared between relatives) for the evolution of altruism and cooperation. In *The Causes of Evolution*, he noted that altruistic behavior "is a kind of Darwinian fitness"[17] and would spread through natural selection "in so far as it makes for the survival of one's descendants and near relations."

The third major figure of the early Synthesis was Sewall Wright (1889–1988), whose contributions stretched late into life.[18] Wright's early career research on guinea pigs in the early 1920s provided key insights as to the effects of inbreeding and led to further studies on the relative contributions of heredity and environment in accounting for variation among individuals. Wright's interest in the inheritance patterns of coat color in these rodents led to a better appreciation of traits controlled by multiple genes, which were not well understood at the time but had clear evolutionary relevance. His theoretical contributions included important concepts such as genetic drift (often termed the "Sewall Wright effect").

Whereas Fisher's quantitative models focused mostly on large populations exhibiting continuous variation, Wright's models incorporated schemes in which populations could be segmented into smaller, partially isolated, fragments.[19] His genetic drift concerns the important recognition that chance fluctuations in allele frequencies could have a significant evolutionary impact. A small, isolated population derived from a large one might not be genetically representative of the larger original one. And the smaller the secondary population, the greater the chance of sampling error, or divergence from the larger population. Wright quantified the effects of genetic drift and showed how it could lead to the fixation of alleles within populations.[20] His ideas contributed to our understanding of how genetic variation is maintained, or lost, within populations.[21]

Ernst Mayr (1904–2005) was an avid childhood bird watcher in his native Germany, had earned his doctorate in ornithology by age twenty-one, and soon after began rigorous and physically demanding field work in unexplored areas of

New Guinea, collecting thousands of avian specimens for later study. He subsequently joined the famous Whitney South Sea Expedition (1920–1941)[22] to collect bird specimens for the American Museum of Natural History, which later hired him as curator of birds. Mayr's field biology experience gave him insights that greatly influenced the Modern Synthesis.

His "biological species concept" addressed an issue that had intrigued and puzzled Darwin, namely, what distinguishes a species? The question had remained unsettled as of the *Scopes* trial. That this problem had not been at all settled at the time of *Scopes* is clear from the expert testimony. Metcalf, for example, stated that "The word species is indefinable, and is used by biologists merely as a convenience, and it has wholly different meanings when applied to different groups of animals and plants."[23] Horatio Hackett Newman echoed that conclusion, stating, "The species is the unit of classification, but there is serious doubt as to whether species have any reality outside of the minds of taxonomists. Certainly, it is extremely difficult, if at all possible, exactly to draw sharp boundary lines between closely similar species."[24]

Mayr argued that species are groups of interbreeding individuals that are reproductively isolated from other groups, a definition of species that became widely accepted in the field of evolutionary biology. He viewed speciation not so much as the origin of new types, but instead as the "origin of effective devices against the in-flow of alien genes into gene pools."[25]

Mayr also made significant contributions to our understanding of speciation, the process by which new species arise. He proposed that most new species arise through "allopatric speciation"[26] in which small populations become isolated around the edges of the original larger group's range. This isolation can lead to genetic divergence and, eventually, the formation of a new species.[27] Mayr's ideas on speciation provided a crucial link between the study of populations and the evolution of new species.

Finally, George Gaylord Simpson (1902–1984), an American paleontologist, played a key role in integrating paleontology—the study of the history of life on Earth based on fossils—into evolutionary biology. Prior to the Modern Synthesis, macroevolution—usually taken as evolution above the species level or evolution of the higher taxa—was largely the domain of paleontology. Few paleontologists accepted that natural selection could account for such phenomena, and thus only a small number of them were strict Darwinians. Many paleontologists accepted soft inheritance and believed that major saltations, rather than gradual changes, were necessary to explain the origin of new species and higher taxa.[28] Simpson helped reverse this using morphological, taxonomic, and distributional evidence to infer that macroevolutionary processes do not require exceptional explanations, but instead are rooted in Darwinian natural selection.

Having joined the American Museum of Natural History as a curator of vertebrate paleontology just after *Scopes*, in 1927, Simpson became intrigued with the evolution of horses. In contrast to the then-standard vision of a straight-line progression in the evolution of modern horses from their ancestral forms, Simpson's novel quantitative approaches to the fossil record revealed a rich branching of early horses over time in diverse ecological habitats as the early populations adapted initially to forests, and then to open grasslands. His work on the fossil record provided crucial evidence for the gradual evolution of species over long periods.

One of Simpson's key contributions was the concept of "tempo and mode" in evolution. He argued that evolution occurs at different rates (tempos) and follows different patterns (modes) in different lineages and at different times. Some lineages experienced relatively rapid evolutionary change, whereas others exhibited more gradual change. This concept helped explain the diversity of life forms and patterns of long periods of stability interspersed with relatively rapid change observed in the fossil record.

Simpson also made significant contributions to our understanding of adaptive radiations, which are periods of rapid diversification of species revealed in the fossil record. (The explosive evolution of mammals after the extinction of dinosaurs is an example.) Simpson's work provided insights into how ecological opportunities can drive the diversification of species. Within a short period after the end of the *Scopes* trial, these remarkable scientists had built a body of knowledge that, taken together, resolved most of the errors and inconsistencies that had plagued evolutionists during the age of Bryan and Darrow.

COMMUNICATING TO A BROADER AUDIENCE

Collectively, the early Synthesists provided a mathematically based case in the early 1930s, showing that even small pressures of natural selection acting on minor genetic differences of Mendelian origin, can result in evolutionary change. The field of theoretical population genetics was born from this work. But these quantitative accounts were challenging and complex, even for scientific audiences (to say nothing of the general public). The second phase of the Synthesis, largely taking place later in the decade, involved several pivotal publications that communicated these innovations in a more accessible way.

Theodosius Dobzhansky (1900–1975) had broad professional and personal interests, and was fluent in six languages—"a man for all seasons."[29] In 1937, Dobzhansky published *Genetics and the Origin of Species*, a pivotal book that brought together the complex and rather abstract theoretical work of Wright, Fisher, and Haldane, presenting it in a way that was comprehensible and compelling to biologists in various subdisciplines that previously had little in common. This was a remarkable achievement, in light of the fact that Dobzhansky himself was not gifted in the mathematics of population genetics.[30] But through his efforts, evolution came to be recognized as that common link across biological subfields.

Genetics and the Origin of Species achieved a systematic overview of relevant experimental population genetics at the time, including Dobzhansky's own work with fruit flies, and it extended the synthesis to speciation and other cardinal problems. Much of the first third of the book looks at the important role chromosomes, especially chromosomal reorganization, can have on speciation. Dobzhansky would later famously claim that "nothing in biology makes sense except in the light of evolution."[31]

Dobzhansky's empirical work included seminal experiments that examined populations of fruit flies in different environments. When populations of the same species were separated and exposed to different environmental conditions, genetic differences accumulated in each. These differences could eventually lead to reproductive isolation, where individuals from different populations could no longer interbreed successfully. This work provided empirical evidence for the role of natural selection and adaptation in driving speciation.

Another writer who played a crucial role in communicating these ideas to a broader audience was, fittingly, Julian Sorrell Huxley (1887–1975), the grandson of Darwin's bulldog. A British biologist, Julian Huxley emphasized the importance of genetics, natural selection, and adaptation in explaining the diversity of life. Huxley had a clear and engaging writing style that made complex scientific concepts more understandable to non-specialists. (The title of his 1942 broad survey book, *Evolution: The Modern Synthesis*, was the origin of the "Modern Synthesis" moniker.) A clever and gifted writer, Huxley coined many terms that gained wide usage, including "ethnic group" (in lieu of race), "clade" (a group of species descended from a single ancestral species), "cline" (a gradual change in allele frequency or a character, like body size, over a geographic transect), and "ritualization" (the evolutionary modification of movements and structures resulting in better signal function).

Ernst Mayr's numerous books on the history of biology and evolution were also significant and influential in spreading the Synthesis (although Mayr has sometimes been criticized for an over-emphasis of his own views and his role in the Modern Synthesis in these accounts). Together, Dobzhansky, Mayr, and Huxley in particular contributed to wider acceptance of the Synthesis by scientific audiences who might not otherwise have had the expertise to engage with the works of Fisher, Wright, and Haldane.

Despite achieving wider acceptance, the Synthesists' work was nevertheless complex and technical. Darwin's tome had not convinced everyone—quite the contrary, as events up to this point demonstrate. But his accounts of natural history, and the observations he made during the voyage of the *Beagle*, were accessible. Most people have had an opinion on evolution, at least since Darwin published *Origin* and perhaps before (since, perhaps, the work of Lamarck, or since Chambers published *Vestiges*). However, the body of evidence supporting the theory has gradually become more complex and less easily conveyed to the general public.

Writing nearly eighty years after *Scopes*, journalist Lauri Lebo documented a court case much like *Scopes*, challenging the Dover, Pennsylvania, school board's decision to remove evolution from classrooms. In doing so, she wrote that, if "Dover's battle had been over quantum physics or (God forbid!) string theory, I would have been hopelessly lost. But evolutionary theory and natural selection are fairly intuitive concepts."[32]

Perhaps. But not long after *Scopes*, the Synthesists had begun amassing evidence for evolution that was (notwithstanding the important efforts of Dobzhansky, Huxley, Mayr, and others) far less intuitive and accessible to the public. For now, as we detail in the next chapter, accounts of evolution had receded from the public eye. But in time, old controversies would return anew to legislatures and classrooms.

CHAPTER TEN

Eugenics, Depression, and the Road to War

> Science has made war so hellish that civilization was about to commit suicide; and now we are told that newly discovered instruments of destruction will make the cruelties of the late war seem trivial in comparison with the cruelties of wars that may come in the future.
>
> —Planned Closing Argument of William Jennings Bryan, Trial of John Scopes

THE 1930S TRANSFORMED THE US LEGAL and political system as much as it had transformed the field of biology. The Great Depression brought new urgency to the progressive struggle to enact economic reforms in the face of conservative judicial resistance. President Franklin Roosevelt's response was the New Deal, an ambitious economic program combining regulatory reforms and public works projects. For the people of Dayton, perhaps the most consequential component would be the Tennessee Valley Authority (TVA), a federal initiative to alleviate poverty in the region by managing and developing the Tennessee River and modernizing the region's infrastructure. The eccentric *Scopes* defense attorney John Randolph Neal championed the program so aggressively that he became known locally as "the father of the TVA."[1] (Characteristically, after the program he fought for was established, Neal spent much of the 1930s criticizing aspects of its administration.)

The New Deal was initially stymied by an inflexibly conservative block of four justices, known in the popular press as the "Four Horsemen," in alliance with relative moderates Owen Roberts and Charles Evan Hughes, the man ACLU leaders had wanted as a replacement for Darrow in the *Scopes* appeal, now returned to the Supreme Court as its chief justice. In this fight, Roosevelt was frustrated and unlucky (he had no opportunities to nominate a justice during his first term). In response, he planned to "pack the Court" by adding additional seats to outvote the opposition.[2]

Court-packing failed politically—but it famously prompted the "switch in time," with Hughes and Roberts bringing their pivotal swing votes to the liberal side, upholding New Deal reforms while preserving the Court's independence. At this moment, the long-running conflict between majoritarianism and the judiciary's old guard, a roadblock in the career of Bryan and a recurring obstacle for the early-century progressive movement, was finally resolved—for the time being, at least. The 1937 case of *West Coast Hotel v. Parrish*[3] is conventionally regarded as marking the end of the *Lochner* era.

Clarence Darrow died the year after *West Coast Hotel*, in 1938, just before he would have turned 81—and just before the outbreak of another war in Europe. Late in life, he and Hays had reunited to work on the American Inquiry Commission, an unofficial organization devoted to documenting and publicizing the atrocities of the Nazi regime.[4]

Those atrocities were rooted in the Nazis' monstrous ideology, which, to some extent, Bryan had presaged in his warnings about the implications of eugenics. As noted earlier, eugenics had been prominent during the *Scopes* era, but neither Bryan nor Darrow supported it. They opposed it, however, from significantly different analytical perspectives. Darrow was skeptical of the science behind its assumptions and proposed applications. He thought it was, for example, naïve to imagine a hereditary component to endemic crime that could be addressed through eugenics and sterilization.

For Bryan, materialistic scientific modernism was fundamentally misguided, and eugenics—including advocacy for voluntary or even mandatory sterilization—was a predictable but unacceptable extension of evolutionary theory. He foretold of "scientific breeding, a system under which a few supposedly superior intellects, self-appointed, would direct the mating and the movements of the mass of mankind."[5] Bryan did not live to see how the views and beliefs of some of the Synthesists in the 1930s and 1940s were not far off from his anxious concerns.

Huxley, for one, had complex views on eugenics that changed over time, especially after World War II.[6] Crafting his words with characteristic care, he used the term "humanism" to describe and promote what he believed was a fair, progressive, and necessary form of eugenics for human survival. He was a longtime member of the Eugenics Society, its vice president from 1937 to 1944, and its president from 1959 to 1962. In the 1930s, he supported the campaigns for sterilization legislation in Britain.[7]

Early in his career, Haldane also believed in eugenics and included in *The Causes*[8] the suggestion that

> [t]he classes which are breeding most rapidly in most human societies today are the unskilled laborers. Society depends as much, or perhaps more, on the skilled manual workers, as on the members of the profession and ruling classes. But it could well spare many of the unskilled . . . However, I have no desire to discourse on eugenics.

Some accounts indicated that Haldane later emerged as a critic of eugenics and that he concluded it was being used for distorted political ends, but there is disagreement on this.[9] His support of eugenics is noteworthy because Haldane was the most clearly left-leaning of the Synthesists, complicating prevailing assumptions that eugenics was strongly associated with the political right. This was reflected, for example, by Haldane's engagement with the Communist Party of Great Britain during the 1940s.[10]

But among all the Synthesists, it was Fisher whose views on eugenics, although sometimes minimized in the literature,[11] most clearly relate to the concerns Bryan expressed at the time of the *Scopes* trial. The darker side to Fisher's reimagining of natural selection and its consequences for humanity is revealed in the last five chapters of *The Genetical Theory* (in particular, "Reproduction in Relation to Social Class," "Social Selection of Fertility," and "Conditions of Permanent Civilization").

These chapters advocated for eugenics. The downfall of civilizations, he claimed, was due to the voluntary decline in fertility of their upper classes.[12] He argued against welfare allowances that enabled large families and was for a time prominent in the Eugenics Society (which had its origins in 1907; one of Darwin's sons, Leonard, served as its second president). Fisher resigned from the Society over what he saw as the decreasing role of science within the movement.[13] Later in his career, Fisher was Galton Professor of Eugenics at University College, London.

Stephen Jay Gould described what he referred to as "a tradition of discreet silence" among some historians of science regarding these awkward chapters of Fisher's classic book.[14] He supposed that this "conspiracy of silence" was likely due to collective embarrassment that such a key figure in the emergence of the Modern Synthesis should have ended his canonical book with such a long argument for a politically discredited movement.[15]

That earlier lack of attention, whatever its motivation, no longer holds, and the field has responded with calls to hold individuals accountable for their actions and beliefs. A recent, direct, and firm criticism of Fisher and his eugenic views details the 2020 decision at Gonville and Caius, a constituent college of the University of Cambridge, to remove a commemorative stained-glass window honoring him.[16] The author, Cambridge Professor Emeritus of History Richard J. Evans, wrote in an article entitled "RA Fisher and the Science of Hatred" that Fisher "was less unsympathetic to Nazi eugenics than most of his British colleagues were" and had campaigned for sterilization legislation for the "mentally defective." Evans argued that memorials "are less about the past than about the present and the future . . . a future that [ought to be] democratic and inclusive," a philosophy that has been widely adopted at colleges and universities evaluating their own naming honors.[17] For years, the R. A. Fisher Prize was awarded annually by the Society for the Study of Evolution for an outstanding Ph.D. dissertation paper published in the journal *Evolution* during a given calendar year. In the summer of 2020, the Society announced its decision to rename the award.

In a recent short but thorough biographical sketch and review, seven biologists and quantitative scientists strove to remind their readers, through a baby-and-bathwater argument, of the many and significant contributions Fisher made over the course of his lifetime.[18] They deemed Fisher one of the greatest scientists of the twentieth century but stressed that their goal in the piece was neither to defend nor attack Fisher's work in eugenics and views on race but "to present a careful account of their substance and nature."[19] They ended with a Hail Mary pass in what they proposed as a constructive path moving beyond the present stage of accountability:

> Rather than dishonouring Fisher for his eugenic ideas, which we believe do not outweigh his enormous contributions to science and through that to humanity, however

> much we might not now agree with them, it is surely more important to learn from the history of the development of ideas on race and eugenics, including Fisher's own scientific work in this area, how we might be more effective in attacking the still widely prevalent racial biases in our society.

In any event, regardless of how one assesses the legacy of the Synthesists today, the support of some of them for the eugenics movement remains a relevant part of the *Scopes* legacy. At a minimum, it demonstrates that some of Bryan's attacks were not as far from the mark as they might have seemed, or as Scopes's defenders might have hoped.

But a few important caveats are in order. First, Bryan and his fellow anti-evolutionists were primarily motivated—like Bishop Wilberforce before them—by perceived contradictions between evolution and literal biblical accounts. The enactment of the Butler Act and the trial of John Scopes was no crusade against eugenics.

Moreover, support for eugenics during the progressive era was widespread throughout society and did not track along easily identifiable lines. Clarence Darrow opposed it wholeheartedly, but many religious figures, including the archbishop of the Diocese of New York, supported it. Indeed, in 1912, in response to the eugenics movement, hundreds of protestant ministers announced that they would not marry couples in their churches without a certificate from a physician certifying their health.[20] Supporters (and opponents) could be found on both the political left and right and among both the religious and the irreligious. Eugenics was hardly an exclusive project of evolutionary biologists, nor was opposition to it an exclusive project of the devout.

Finally, care with this subject is especially warranted because both real and exaggerated connections to eugenics during the Progressive era have been disingenuously used as a cudgel against a variety of modern movements with no contemporary ties to eugenics, merely because they existed at a time when eugenics had gained some broader social acceptance. Opponents of abortion, for example, have often charged that early leaders of Planned Parenthood, including founder Margaret Sanger, endorsed eugenics, in a misleading effort to equate reproductive healthcare with eugenics, sterilization, and even Nazism.[21] But modern abortion activists, providers of reproductive health, or for that matter, evolutionary biologists, have no more connection to early-century eugenicists than modern religious organizations do.

Notwithstanding these important qualifications, Bryan's damning charge, in his post-trial speech, that evolutionists' program for humanity was a system of "scientific breeding" indeed has disturbing echoes in some of the subsequent writings of Fisher, Haldane, and Huxley. Regardless of their intentions, the fact that Bryan's attacks contained a kernel of validity ultimately provided ammunition for those who would continue his fight long after the Synthesis had resolved most lingering scientific questions about evolutionary mechanics.

Highly exaggerated links between evolution and eugenics have become a cornerstone of anti-evolutionary rhetoric: a typical film attacking evolution[22] featured a speaker discussing simplified and overstated connections between Darwin and the eugenics movement, filmed with hyperbolic insensitivity at the site of a Nazi death camp. Likewise, in the early 2000s, Democratic state representatives in Georgia introduced a resolution condemning the state's historical involvement with eugenics and mandatory sterilization—a Republican state senator allowed the measure to proceed only on the condition that background to the resolution be added condemning eugenics as an "outgrowth of Darwinian evolutionary theory."[23]

Neither R. A. Fisher nor Francis Galton are household names the way Charles Darwin is, and in these crude attacks, Charles Darwin himself is often inserted in the place of those who actually advocated for eugenics. The result has been an unfortunately widespread popular belief that crudely equates Darwin with racism. (Captain FitzRoy, who nearly banished the young Darwin after he bitingly argued against slavery, might be surprised.) The impression still lingers. In a 2005 court decision, for example, Judge Sam Lindsay, a federal judge in Texas, wrote:

> [A]s part of Darwin's theory, he believed, and many thought he had proved, that whites evolved from chimpanzees, which were considered the most intelligent of the apes; that Asians and brown-skinned persons developed from orangutans; and that blacks evolved from gorillas, which were the least intelligent and civilized, but the strongest and most brutal and savage of the apes.[24]

None of this was a part of Darwin's work, and Lindsay offered no support for his argument through citations to Darwin's writing or any other source. (On the contrary, as discussed already, Darwin, along with his friend and chief American advocate Asa Gray, had asserted the shared evolutionary ancestry of all humans, whereas Gray's opponent, the anti-evolutionist scientist Louis Agassiz, had argued for the separate creation of human races.[25]) This poorly informed passage from Lindsay's opinion illustrates how these impressions linger, not just in fringe polemics but even in official documents.

Once again, Darrow opposed eugenics no less than Bryan did, as did many others who held a materialistic and scientific worldview—and as do all reputable evolutionary biologists today. The fact that a few of Bryan's attacks ended up disturbingly close to the mark remains a notable, if uncomfortable, piece of the *Scopes* legacy with continuing ramifications.

In the following chapter, we follow the end of the Second World War through the beginning of the Cold War—an event that, perhaps surprisingly, brought evolutionary science back to US classrooms and courtrooms. The years following *Scopes* proved to be a brief interlude, and the 1950s would be the last decade to pass without a *Scopes*-style legal case about the place of evolution in public school classrooms.

CHAPTER ELEVEN

The Midcentury Moment

[I]n no possible way can science be taught or science be studied without bringing in the doctrine of evolution.

—Argument of John Randolph Neal, Trial of John Scopes, Day 2

THE SECOND WORLD WAR CONSUMED nearly the entire planet. For the general public, little about the evolution debates of prior decades seemed comparably important. When noted author and foreign correspondent William L. Shirer, who reported from Europe during the war, looked back on the 1920s, he merely recalled his desire to flee "the arid land of Coolidge, Prohibition, and the Scopes 'monkey trial,'" which he considered "an example of the inanity of American life in that silly time."[1] Few of the surviving *Scopes* participants played a major role in the war, but George Rappleyea, the man who had brought the case to Dayton, used his engineering skills to produce landing craft. He did so at a New Orleans–based company, having left the small Tennessee town he had once brought to the nation's attention.[2]

Evolutionary science had profoundly interested elite legal thinkers during the decades leading to *Scopes*. As already recounted, Holmes's writings were replete with references to ideas stemming from Darwin and Spencer. Legendary contracts scholar Arthur Linton Corbin wrote in 1914 that in "the judicial world, as in the animal and vegetable world, the ultimate law is the law of the survival of the fittest."[3] The great evidence scholar John Henry Wigmore helped develop an elaborate evolutionary model for the law between 1915 and 1918.[4] But like the public's interest in *Scopes*, references to evolution in US legal writings faded between the trial and a later revival in the 1970s.[5] It was as if *Scopes* had, temporarily, exhausted the interest of non-specialists in a question that had, up to that point, consumed public debate.

Soon after the war, the unexpected collapse of the US-allied nationalist government of China ushered in a new era defined by pervasive fear of communism.[6] Both houses of Congress launched intrusive investigations into alleged subversive activities—some real and many imagined—across a range of industries, cultural institutions, and government offices. Bills criminalizing membership in communist organizations were swiftly enacted. Permanent residents were deported for having only tenuous former connections to communist political movements.[7] The "second Red Scare" had begun, and it would soon outpace the first.[8] (One surprising figure to inform on left-wing radicals before the anti-communist House Un-American Activities Committee was none other than Benjamin Gitlow, the communist radical Darrow had once defended, whose case had marked a turning point in First Amendment jurisprudence. Having served time for criminal anarchy, Gitlow pivoted to a second act: a successful career as a "professional informant.")[9]

Two discontents were playwrights Jerome Lawrence and Robert Lee, who reached back to *Scopes* for inspiration and analogy. In doing so, they brought the events of 1925 back into the public eye. Their highly successful 1955 drama *Inherit the Wind* played a pivotal role in constructing the historical memory of the trial. Its fictionalized version of Bryan, a demagogue named Matthew Brady, serves as the play's stand-in for anti-communist crusader Joseph McCarthy—and for all intransigent defenders of rigid orthodoxy.

Although *Inherit the Wind* was not the first play to grapple with the cultural undercurrents of the Red Scare, it differed from previous works in its focus on one man's (not entirely successful) demagoguery rather than majoritarian ill-treatment of a targeted central character. Unlike many other liberal responses to the McCarthy era, *Inherit the Wind* is not a depiction of a martyr to the prejudices of the mob. Rather, the play depicts a powerful but ignorant avatar of conservatism, railing against "Bible-haters" and "brewers of poison," trying—and, significantly, *failing*—to stand against the tide of progress.[10]

After achieving financial success and running for more than eight hundred performances, the play's first run closed on Broadway in 1957. It would spawn many award-winning revivals, television productions, and one feature film, definitively stamping its version of the *Scopes* trial into American legend. This was a version defined by the duel between imagined versions of Bryan and Darrow, with other players like Hays and Bryan Jr. largely forgotten. And it made its mark. Only after the play debuted on Broadway did the real trial receive an entry in *Encyclopedia Britannica.*[11]

Cultural developments of the postwar era had a profound impact on the scientific community as well. Historians such as Daniel J. Kevles contend that eugenics was repudiated after the Nazis' atrocities became widely known.[12] Perhaps relatedly, in a decision issued during the war, the Supreme Court reached a contrary result from *Buck v. Bell,* striking down an Oklahoma law that provided for compulsory sterilization of felons.[13]

That said, supporters of eugenics did not completely disappear, as is clear in the later writings of R. A. Fisher and other Synthesists. Julian Huxley continued to believe, as late as 1962, that

> Eugenics can make an important contribution to man's further evolution: but it can only do so if it considers itself as one branch of that new nascent science [of Evolutionary Possibilities], and fearlessly explores all the possibilities that are open to it.[14]

To guard against the threat of genetic deterioration from nuclear fallout, Huxley went on[15] to advocate the building of subterranean shelters for sperm banks that should include collections of deep-frozen sperm from "a representative sample of healthy and intelligent males" (evoking a scene from Stanley Kubrick's 1964 classic satirical film, *Dr. Stangelove*).

A popular evolutionary biology college textbook from the late 1970s still included content on eugenics, although the authors also included a section (titled "Should Mankind Steer Its Own Evolution?") devoted to ethical and philosophical considerations of the material.[16] (The text also included an uncritical description of Nobel laureate H. J. Muller's proposal to create sperm banks for the seminal fluid of men of great achievement.)

Accordingly, the exact moment when eugenics became a tainted concept, even a slur—a word irretrievably linked to oppression, discrimination, and human rights violations—has been the subject of debate.[17] But without doubt, the atrocities of the Nazi regime did much to discredit the eugenics movement. After Germany's defeat and revelations emerged about the horrors it had inflicted across Europe, ideas that even vaguely resembled those of the Nazis faced heightened scrutiny.

The Modern Synthesis resurrected Darwinian natural selection and, while doing so, expanded evolutionary thinking to points and places Darwin never imagined. Emboldened by their successes, the Synthesists—some of them, at least—assumed a premature responsibility and imagined a path for human destiny that was inconsistent with the sensibilities and ethics of the postwar era—to say nothing of our modern values. How history views their motivations and, overarchingly, assesses and remembers their lifetime contributions is a subject for deep reflection and scholarship.

But accountability, in science and elsewhere, can be tricky and will often rest in the eye of the beholder, and the stakeholder. Indeed, Lawrence and Lee captured something of this sentiment in the final scene of *Inherit the Wind*. After the court clears, the sudden death of the Bryan character, Matthew Harrison Brady, is announced and the reporter (a character based on Mencken) shows no sorrow: "leave the lamentations to the illiterate." But Drummond (Darrow), to Hornbeck's apparent surprise, magnanimously comes to Brady's defense, refusing, he says, "to erase a man's lifetime." Brady, he says, "had the right to be wrong," and continued, "A giant once lived in that body. But Matt Brady got lost. Because he was looking for God too high up and too far away."

THE RETURN OF *SCOPES* TO THE NATIONAL DEBATE

After *Inherit the Wind*, public interest grew in reassessments of the trial—but few questioned the conclusion that, even though *Scopes*-era laws like the Butler Act remained on the books in three hold-out states, evolution had won the great debate. Fay Cooper-Cole, the pioneering anthropologist who had submitted testimony in the *Scopes* defense, recounted his experiences in the *Scientific American* in 1959. In his assessment, no "attempt at repression has ever backfired so impressively," as the *Scopes* prosecution.[18]

The following year, the *University of Chicago Law Review* featured a special discussion of the 1925 case, perhaps inspired by *Inherit the Wind* (which was referenced in the review's introductory paragraph). The volume featured contributions by legal scholars Thomas Emerson, David Haber, Malcolm Sharp, and

Harry Kalven Jr.[19] True to the spirit of the age and the influence of the play, these academics analogized *Scopes* to debates over communism. As for biology, Emerson and Haber concluded that, in their "more sophisticated" age, it was "unlikely that any religious group could muster political power for an outright suppression of the teaching of science." Indeed, they went so far as to state that the "specific doctrine of evolution, at least in its general form, is now as thoroughly established in popular thinking as the Copernican theory of the solar system," and that it would be "inconceivable" for a state legislature to attempt to ban it.[20] (Subsequent history reveals otherwise.)

But it was an unrelated Cold War milestone, taking place just months after *Inherit the Wind* ended its first run, that set off a chain of events that would eventually thrust evolution back into US courtrooms. In October 1957, the Soviet Union launched a rocket carrying an object named "artificial fellow traveler around the Earth," or, shortened and in Russian, *Sputnik*. Humanity's first artificial satellite orbited the planet for three weeks, completing an orbit every ninety minutes and broadcasting a radio signal back to the surface. Sputnik appeared to give the USSR a commanding early lead in the budding space race. Up to that point, the US public and many policymakers had derided Soviet technical capacity as lagging far behind. Now, media and legislators railed against the Eisenhower Administration for failing to keep up, pointing to perceived weaknesses in American scientific training. Some liberals even suggested that the chilling investigations of the McCarthy era had damaged the careers of American scientists who might otherwise have contributed.[21]

One consequence was a redoubling of investment in American science. The National Defense Education Act flooded money to US schools, contributing to what would become a golden age of science in public classrooms. The year after Sputnik, the National Science Foundation began funding several programs aimed at jump-starting scientific education, including grants for an organization—the Biological Science Curriculum Study (BSCS), led by Johns Hopkins geneticist Bentley Glass and University of Florida zoologist Arnold Grobman—created to modernize US schools. Under the leadership of Glass and Grobman, the BSCS strongly endorsed updates to biology curricula, which would inevitably put greater emphasis on instruction about evolution.

Today, there is some debate about whether evolution was still being taught in U.S. high schools before the intervention of the BSCS. The conventional account, relayed in several articles and recounted as background in a significant federal court decision, is that the *Scopes* trial motivated high school biology textbooks to remove, or at least significantly deemphasize, evolution for fear of continued controversy. BSCS reversed this trend, putting evolution back into classrooms and teeing up the renewed debates of subsequent decades.[22] That account has been challenged by Ronald Ladouceur and others. Ladouceur has argued that the success of anti-evolutionists in driving evolution out of textbooks in the period between *Scopes* and *Sputnik* has been exaggerated by earlier works.[23]

Either way, modernized biology materials, which emphasized an updated account of biological science, guaranteed greater prominence for evolution in the 1960s. These updated materials would be informed by the Modern Synthesis and were thus based on more stable and accurate (though complex) explanations for evolutionary mechanisms than were available in the 1920s. These phenomena set the stage for legal conflict in the three states that retained *Scopes*-era anti-evolution laws: Arkansas, Mississippi, and, of course, Tennessee.[24]

The year was 1967. Five years had passed since Spencer Tracey earned an Oscar nomination for portraying Darrow ("Drummond") in *Inherit the Wind*. *Scopes* (or, at least, a dramatized version of it) had returned to the popular imagination, and Tennessee science teacher Gary Scott initiated a lawsuit challenging the venerable Butler Act. Rather than risk an embarrassing repetition of the *Scopes* trial, the state legislature simply repealed the law.[25] Only two states remained.

THE TRIUMPH OF EPPERSON

The fight came next to Arkansas and would lead to a through-the-looking-glass case that would play out as *Scopes* in reverse. In the wake of the *Scopes* trial, Arkansas adopted an anti-evolution statute modeled on Tennessee's. Arkansas's law forbade instructors at any state educational institution to teach, or select textbooks teaching, "the theory or doctrine that mankind ascended or descended from a lower order of animals." Unlike the Butler Act, Arkansas's law had been adopted by referendum in 1928, with the people voting it into law by a margin of 108,991 to 63,406.[26]

Arkansas schools had later adopted biology textbooks containing sections on Darwinian evolution—and the use of those official textbooks was technically criminal under the longstanding but moribund 1928 law. An Arkansas reporter tracked down the aging John Scopes to inquire about the issue. Scopes responded that such laws, once they were on the books, were difficult to repeal through the political process.[27] In that, he was right. But might these remaining laws be struck down by a federal judiciary that had been transformed and emboldened since the fall of *Lochner*?

The legal landscape had shifted in many ways during the intervening years. Significantly, the First Amendment to the U.S. Constitution, with its protections for free speech and religious liberty, was now understood to apply directly to the states, rather than only to the federal government. This was the "incorporation" doctrine that Holmes had presaged in the *Gitlow* case, now accepted by a majority of a transformed Supreme Court, guided by a liberal and ambitious majority.

A Little Rock teacher named Susan Epperson, together with a parent of a student, sued in state court seeking a ruling overturning the anti-evolution law. Epperson never achieved the fame that *Scopes* did, largely because her case was far less dramatic. It featured no legendary duels between the likes of Darrow and Bryan. Significantly, due to intervening legal developments, Epperson was able to bring her case *offensively*, suing to challenge the law without incurring any legal jeopardy herself. Unlike *Scopes*, her team did not need to arrange for her arrest, and the proceedings would be civil, not criminal.

In the Arkansas popular press, a reporter noted that the state's anti-evolution law meant different things to fundamentalists, liberals, and educators. To the first group, it was religious symbolism; to the second, "darkness and witches and monkey trials," and to the third, an affront to academic freedom.[28] In terms of legal strategy, it still remained an open question whether Epperson's case would be seen primarily in terms of the speech rights of a teacher or as a matter of the separation of church and state. The most thorough academic treatment of the question at the time, Emerson and Haber's review of the *Scopes* case for the *Chicago Law Review*'s retrospective, argued that such a case was best seen as presenting a free-speech problem.[29] Like Darrow and Hays before them, Epperson's attorneys chose to advance both a freedom of speech theory and a religious establishment theory, not knowing which, if either, would ultimately persuade the judges.

Initially, her suit met with success. Although the state's elected attorney general planned to use the lawsuit to gain publicity—likely seeking to imitate the more famous *Scopes* trial—he was no William Jennings Bryan. Indeed, he was not even a Tom Stewart. According to a later interview with Epperson, he echoed Bryan by asking her on the witness stand whether she had studied the teachings of Friedrich Nietzsche—but he mispronounced the German philosopher's name to such an extent that a local reporter sarcastically asked if he meant Green Bay Packers linebacker, Ray Nitschke.[30]

But although the chancery court ruled in favor of the plaintiffs, the state supreme court overturned that decision in a two-sentence per curium decision. (One justice even objected to the inclusion of the second sentence!)[31] Justice Harlan of the U.S. Supreme Court decried the state court's terse opinion as "deplorable," accusing it of simply passing the buck on a controversial case.[32] Epperson's team sought Supreme Court review.

In this fashion, a *Scopes*-era evolution law finally reached the nation's highest court. Even this was not without controversy. After her trial-court testimony, Epperson's husband, an Air Force captain, had been assigned to the Pentagon. The family moved to the Washington, D.C. area (and even attended oral argument at the Supreme Court "as tourists" in their own case).[33] Strictly speaking, this meant that the case no longer presented a case or controversy for the Court to decide—a doctrine known as "mootness" would normally mandate its dismissal.[34] But most of the justices seemed eager to decide the case and studiously ignored its procedural problems. As for Arkansas, new state attorneys had been elected, and unlike their predecessor, they had no interest in imitating Bryan. The Court had scheduled an hour for arguments, but they ended after only thirty-five minutes when the state had little to say in defense of its own law.

The resulting majority opinion was crafted by Justice Abe Fortas, who came to the Court in 1966 with a reputation as a successful liberal attorney. In some ways he resembled Darrow—Fortas won early attention by skillfully defending those accused of disloyalty during the McCarthy investigations.[35] Fortas was a friend and ally of President Lyndon Johnson, who had cleared a Supreme Court vacancy by naming sitting Justice Arthur Goldberg ambassador to the United Nations.

In 1968, the year *Epperson* was decided, Johnson re-nominated Fortas, seeking to elevate his friend from associate to chief justice upon the retirement of Earl Warren.

That was not to be. Fortas would be subjected to the first filibuster of a Supreme Court nomination when a conservative block in the Senate preferred to allow the incoming Nixon, rather than the outgoing Johnson, to select the next chief. During the ensuing controversy, a scandal arose relating to a former client's lifetime retainer, from whom Fortas continued to receive funds, and he resigned in 1969 having served fewer than four years on the Court.[36] A bold and intelligent jurist, Fortas's influence would likely have been significant had his tenure not been cut short.

Nevertheless, he left a real mark one year before his resignation, in *Epperson v. Arkansas*. Naturally, Fortas began with a reference to *Scopes*, noting that the challenged law was modeled on "the famous Tennessee 'monkey law' . . . upheld by the Tennessee Supreme Court in the celebrated Scopes case in 1927." He also quoted a sarcastic remark by Darrow, who had pointed out that states that banned evolution continued to teach that the Earth was round. Fortas's assessment of the *Scopes* trial is clear from the beginning of the *Epperson* opinion, and he stood firmly on the side of Darrow.

Fortas unabashedly embraced the theory that Arkansas's anti-evolution law impermissibly violated the separation of church and state. He explained that the government must be neutral toward religion—it may neither aid religion nor exhibit hostility toward it. Fortas also looked to the purpose of the challenged law. Although the text of Arkansas's law was less explicit about its purpose than the Butler Act had been, Fortas nonetheless explained that Arkansas had sought to "blot out a particular theory because of its supposed conflict with the Biblical account, literally read." This was plainly contrary to the First Amendment. The *Epperson* opinion would be hugely influential in future cases involving the separation of government and religion—and it clarified that these evolution disputes indeed implicated the relationship between religion and government. Just as critical was the theory Fortas had *not* relied on: the idea that speech concerning evolution implicated a teacher's freedom of expression in the classroom.

No one dissented, but a negative response came in a concurrence by the venerable Justice Hugo Black. Black was the first of eight Supreme Court justices appointed by Franklin Roosevelt (he outlasted all but one). Before his appointment, Black had been a Democratic senator and supporter of the New Deal; Roosevelt had a justified confidence that, on the Court, Black would help break the opposition of the Four Horsemen. As it happened, Black arrived just after *West Coast Hotel*, the case marking the beginning of the end of the *Lochner* era and the Court's shift toward accommodation with legislative reforms. Black would later write, in *Ferguson v. Skrupa* (and echoing Holmes in *Lochner*), that whether "the legislature takes for its textbook Adam Smith, Herbert Spencer, Lord Keynes, or some other is no concern of ours."[37]

Black was a man of contradictions. He had been involved with the Ku Klux Klan in his youth, but for years as a justice, he was regarded as a leader of the Court's liberal block, playing a significant role in arguing that the Bill of Rights safeguarded civil rights against state governments.[38] Although in Bryan's era, a major part of the ideal of a liberal justice had been a willingness to *defer* to legislatures, under the leadership of Earl Warren, the Court had shifted toward a different strain of liberalism, one equally focused on rights protecting individuals *against* majorities.[39] This shift occurred while Black was on the Court. Nevertheless, at this late stage of his career, Black was simply unpredictable. In the words of judge and scholar

Richard Posner, he had come to display a "quirky populist streak and an autodidact's dogmatism."[40]

Black was the one justice to focus on the standing problems presented by Epperson's case, lambasting Arkansas's "pallid" defense of its own law and questioning whether the case presented a real controversy between the parties. And while he conceded that the law might be too vague to validly enforce, he criticized Fortas for holding that it violated the First Amendment's religion clauses—and in doing so expressed his own doubts about evolution. Black wrote:

> Certainly the Darwinian theory . . . is not above challenge. In fact the Darwinian theory has not merely been criticized by religionists but by scientists, and perhaps no scientist would be willing to take an oath and swear that everything announced in the Darwinian theory is unquestionably true.

Black also questioned whether the Court's decision "infringes the religious freedom of those who consider evolution an anti-religious doctrine."[41]

Hugo Black's concurring opinion illustrates the limits of the influence that the Modern Synthesis had achieved. His statement that "Darwinian theory" had been criticized by scientists, to the point that "no scientist" would take an oath affirming all of Darwin's ideas, quite accurately reflected the state of the field in 1925—when natural selection was in decline, sexual selection had yet to catch on at all, and alternative ideas like mutationism were ascendant. But Black's statement certainly did *not* reflect the state of the field after the Modern Synthesis, when his concurrence was written. Perhaps judicial understandings of science will typically lag; after all, judges are almost never scientifically trained.[42] Even though Dobzhansky, Huxley, and Mayr had done much to explain the conclusions of the Synthesists to a wider audience, outside of the scientific community, misunderstandings remained, even at elite levels—including the nation's highest court.

Of course, there remained many unanswered scientific questions, especially about evolutionary processes in natural populations, and the revolutionary world of evolutionary molecular genetics was only just being discovered. Still, by this point in time, the biological field had embraced a broad, general acceptance of the Synthesis, and a vast accumulation of evidence on many fronts supported it. Nevertheless, there remained surprising levels of resistance. The challenge to Darwinism is often summarized, historically, as having been largely settled by the 1960s. For the evolutionary biologists, this was largely so, but not everyone agreed with them.

The year before Fortas handed down his *Epperson* decision, Ernst Mayr attended a conference at the Wistar Institute[43] in Philadelphia at which evolutionary biologists, physical scientists, and mathematicians assembled to debate the question of whether 4.5 billion years was long enough for the diversity and adaptation seen in the world to have evolved. Yes, this was essentially the same basic question that Darwin and his supporters had to contend with a century earlier, from Kelvin and others. In 1967, the accepted time frame was vastly larger, yet it was still debated. The physical scientists emphatically said "No!" They argued that 4.5 billion years was nowhere near sufficient. Not surprisingly, the evolutionary biologists wholeheartedly disagreed. The two sides went back and forth on the matter for three days and then went their separate ways with nothing resolved.[44]

After the conference, Mayr tried to figure out why there had been such striking disagreement over this key issue. It dawned on him, stemming from some of the things said during the sessions, that the physical scientists had apparently assumed that all individuals in a species were essentially identical, somewhat like the atoms in a mineral. He suspected, as a result, that they believed mutations had to first originate and then spread to *all* individuals, a process that would then be followed by another mutation with the same trajectory, and on and on. This would indeed take huge amounts of time, as Mayr fully appreciated.

But that is not how selection works. Mayr understood that selection operated on many genes in concert, with the individual organism as the unit of selection. In his view, it is most impactful on small populations of individuals that can rapidly give rise to new species. This was cornerstone thinking in the Modern Synthesis, so why did these foundational concepts appear elusive to the physical scientists?

Some of the blame, Mayr contended, had to do with enduring, but severely limiting, conceptions on the part of some of his biological colleagues—he thought the geneticists in particular—who still defined and thought of evolution in the over-simplified terms of changes in gene frequencies. Mayr argued that this parsimonious but constraining conceptualization was insufficient for understanding populations of individuals in the real world, even if it did suffice for closed populations of fruit flies living in a laboratory bottle, or the abstract hypothetical alleles that were the evolutionary entities explored in the models of the theoretical population geneticists. A solution required more work on natural populations, Mayr claimed. Important work in that area, however, was already being performed.

There was at this point, for a brief moment, the appearance that all *Scopes* controversies—and virtually anything resembling them—had been resolved. A decisive rejection of anti-evolution laws seemed to eliminate any legal questions. Scientifically, the Synthesis had woven together a complex and intricate network of supporting evidence that answered most of the questions that had divided the evolutionary field in 1925. Understanding of those developments was not universal, as both Black's concurrence and the Wistar conference made clear, but it was spreading, slowly but surely. Malign misapplications of Darwinism, particularly eugenics, had been discredited and were in retreat. Bryan had briefly triumphed before a Dayton jury, but Darrow seemed to have won in the end—just as Lawrence and Lee had depicted.

CHAPTER TWELVE

Lemon and Peppered Moths

Are we to have our children know nothing about science except what the church says they shall know?

—Argument of Dudley Field Malone, Trial of John Scopes, Day 5

MAYR MAY HAVE RETAINED discontents about broader acceptance of the synthesis, but supporters of evolution had, at this point, achieved all they could have hoped for in the courts. Arkansas's law was now invalid. And with *Epperson* standing as an unambiguous and controlling precedent, Mississippi's Supreme Court was forced to strike down the final *Scopes*-era anti-evolution law in 1970.[1]

One year after the fall of Mississippi's evolution ban, the Supreme Court entrenched Fortas's reasoning and issued one of its most influential decisions relating to the separation of church and state in *Lemon v. Kurtzman*.[2] That case involved lawsuits challenging public aid to private religious schools. The Court struck down these programs, holding that they violated the First Amendment's prohibition against the establishment of religion.

Lemon was transformative due to its articulation of a clear, direct, and straightforward mechanism for resolving cases implicating the separation of church and state—an area of the law that had, up to that point, been relatively nebulous. *Lemon* set forth a three-part test: For a law to comply with the Constitution's directive against governmental establishment of religion, it must (1) have a secular legislative purpose, (2) have a primary effect that neither advances nor inhibits religion, and (3) not foster excessive government entanglement with religion. Fortas was already gone from the Court, and the majority did not directly cite his opinion. But the *Lemon* test clearly echoed his reasoning in *Epperson*—especially Fortas's focus on the impermissible religious purpose of Arkansas's anti-evolution law and his explanation that government may neither aid nor impede religion.

Lemon was not explicitly an evolution case, but it became (for a time) the definitive standard for nearly all cases involving the separation of church and state. Accordingly, a striking number of cases involving evolution in schools were decided in the following decades through the application of the three-part *Lemon* test.[3] Furthermore, when several state legislatures referred potential anti-evolution bills to their state's attorney general to evaluate their legality, the resulting opinions regularly applied the *Lemon* test and concluded that the referred bills were impermissible,[4] thus discouraging the enactment of further bans. *Lemon* had dealt a major blow to the anti-evolution movement.

In the wake of multiple defeats at the hands of a liberal Supreme Court, anti-evolutionists quickly pivoted away from advocating for *Scopes*-era criminal prohibitions against teaching evolution. Different, indirect approaches initially seemed to hold some promise. A few minor court decisions after *Epperson* speculated that requiring *balance* between evolution and creationism might be a constitutionally permissible compromise.[5]

One alternative path that anti-evolutionists explored was to advance their own version of the "academic freedom" argument that the defense had raised at *Scopes*. The postwar Supreme

Court, with its greater emphasis on individual freedoms, had largely abandoned Holmesian greater-includes-the-lesser reasoning as applied to government employees. For example, the same year it decided *Epperson,* the Court had held that a school district could not necessarily fire a school teacher for exercising his First Amendment speech rights—even when the speech in question was criticism of the school board.[6] Holmes's maxim that a policeman "may have a constitutional right to talk politics, but he has no constitutional right to be a policeman" was cast aside. Instead, teachers were no longer expected to "shed their constitutional rights to freedom of speech or expression at the schoolhouse gate."[7]

On the surface, this new focus on the expressive rights of teachers suggested that although a state might no longer be able to discipline a teacher for *expressing* support for evolution, perhaps by the same token it could not discipline a teacher for *rejecting* evolution, or for choosing to teach the Bible in a science classroom.

But even with a more flexible strategy and some favorable legal developments, midcentury anti-evolutionists still faced two insurmountable legal problems. First, Fortas had not pinned his *Epperson* opinion on a teacher's freedom of expression. Instead (over Black's objections), he focused on the religious purpose of Arkansas's law and the resulting impermissible relationship between government and religion. Had Fortas instead accepted alternative arguments structured around a teacher's speech rights—arguments that had been favored by Baldwin in 1925—later anti-evolutionist appeals to teachers' freedoms might have been far more successful. But given Fortas's focus on religion rather than speech, it would be relatively easy for defenders of evolution to demonstrate that opponents acted for a religious purpose and thus fell afoul of *Epperson.*

The second problem anti-evolutionists faced was *Lemon.* By distilling church-and-state separation principles into a straightforward and generally applicable test, the Supreme Court had made the First Amendment's Establishment Clause easier for state and lower courts to apply consistently. A sequence of decisions from many different courts throughout the 1970s, 1980s, and 1990s applied the *Lemon* test to evolution cases. Generally speaking, different courts often apply the same body of law to similar facts and nonetheless reach different results—some variation between judges is inevitable. But from the 1970s through the 1990s, across the board, the application of the generally applicable *Lemon* test, bolstered by the more specific *Epperson* precedent, produced consistent results in evolution cases. The anti-evolutionist response to *Epperson* was thus met with an unmitigated string of defeats for decades.

The final phase of the Synthesis also achieved a triumph: an experiment that would become a classic (though one later saddled with unnecessary controversy). Adaptation is a core tenet of Darwinian natural selection, and the concept itself is quite straightforward: an adaptive trait is one that enhances fitness relative to alternative traits. Yet adaptation has also been described[8] as "a special and onerous concept that should not be used unnecessarily."[9] Identifying adaptations and providing adequate empirical support for them is challenging.[10] Counting fruit flies in the lab is one thing, but measuring natural selection in the wild requires even more creativity.

One of the earliest and most famous field experiments to test natural selection under natural conditions examined the evolution of melanism in the peppered moth (*Biston betularia*).[11] A trend had been noted since the early nineteenth century: in industrialized areas of Britain, the peppered moth, once primarily made up of light gray individuals, now consisted primarily of dark gray individuals. Why had this shift in the moth population occurred? And why were dark gray moths more common in industrial areas whereas light gray moths remained predominant in rural areas? Some persistent theories at the time postulated a direct effect of the pollutants on the moth's color.

Famous peppered moth experiments would be performed by Bernard Kettlewell, who had an impressive academic lineage. Kettlewell was supervised by Edmund Brisco (Henry) Ford, a pioneer in the new area of ecological geneticists.[12] Ford, in turn, had been taught genetics by Julian Huxley at Oxford and maintained a productive and long collaboration with R. A. Fisher. (On the other hand, Ford and Haldane, it is said, did not get along all that well, partially due to their larger-than-life and contrasting personalities and world views.[13])

Kettlewell's experimental approach, which began in the 1950s with complementary work and confirmation and continued into the 1980s, involved placing (dead) moths of the two color morphs onto tree trunks in both polluted and unpolluted areas to study the predation rates by birds. He hypothesized that melanic moths had a selective advantage in polluted environments because their dark coloration made them less easily spotted by birds. Kettlewell employed a "mark-release-recapture" experiment to test this (moths that were spotted and eaten by birds would, of course, not be recaptured). The experiment done in the polluted Birmingham wood resulted in the recapture of 27.5% of dark moths and 13.1% of light-colored moths, a two-to-one advantage for the melanic form. In an unpolluted wood, the recapture rate for light moths was 13.7%—3 times that for the melanic form, for which only 4.7% were recaptured.[14]

Kettlewell's experiments—especially as they were presented in popular magazines and venues[15]—provided compelling and intuitively convincing evidence that selective predation by birds influenced the frequency of melanic moths in industrialized regions. They supported the concept of natural selection acting on heritable traits, a central tenet of Darwinian theory, and the study became widely known just in time for the celebration of the hundredth anniversary of the publication of *Origin*. It also became a ubiquitous textbook case study for the topic of evolution (usually presented in a highly distilled form that does not capture the richness and complexity of the phenomenon), and the study inspired much further research on the topic.

However, the peppered moth study later came under attack. In 2002, journalist Judith Hooper published a book that ignited considerable debate around Kettlewell's classic experiments.[16] The book itself, highly controversial in its claims, did not receive positive reviews in scientific journals and was described as having a scandal-mongering tone and promoting inaccurate assertions and assumptions.[17] Hooper contended that Kettlewell's experiments had not been well-designed, accused him of being a sloppy researcher, claimed that the study's data were suspicious—perhaps even partially fabricated—and argued that there was a broad conspiracy in the Ford research group to hide all of this. She wrote in the book's prologue, "For the record, I am not a creationist, but to be uncritical about science is to make it into a dogma."

Although this sounded virtuous, the science behind the questions Kettlewell and Ford pursued had been replicated and complemented many times by different researchers,[18] and extensive and detailed rebuttals to Hooper's claims quickly appeared.[19] Nevertheless, the mini-scandal that followed Hooper's publication illustrates that many concepts in evolutionary biology can be challenging—adaptation prime among them—even when the evidence *appears* straightforward.

Underscoring this point, consider the trajectory of prominent evolutionary biologist Jerry Coyne's thoughts regarding the Kettlewell story. One might assume that Coyne's best-selling 2010 book, *Why Evolution is True*,[20] would have prominently featured the classic peppered moth study, but for a time, Coyne had reservations about the original moth research, arguing that "in evolutionary biology there is little payoff in repeating other people's experiments, and . . . our field is not self-correcting because few studies depend on the accuracy of earlier ones."[21]

Contrary to Coyne's expression of concern about the lack of self-correction in the field of evolutionary biology, the peppered moth experiment presents a clear example of how the scientific process works to accomplish this goal in evolutionary studies. With the accumulation of supporting evidence, Coyne reversed his earlier assessment and now endorses Kettlewell's original conclusions.[22] However, Coyne's question about why there was general, largely unquestioned, and premature acceptance of the original study's conclusions remains pertinent, because it applies equally well to a host of other more recent examples where evidence for adaptation is less than ideal—especially when it comes to claims about human nature and behavior.

Despite the brilliant success of the peppered moth experiment, all was not entirely stable within the field. Stephen Jay Gould later characterized this as a "hardening" of the Modern Synthesis that began in the mid-1960s.[23] By this, he meant a seemingly self-satisfied, even sanctimonious, reluctance to entertain or embrace alternate views, or even to expand the core ideas that had defined the Synthesis to that point.[24] Gould claimed that the pluralism and openness to novel ideas that had characterized the early years of the Synthesis had been lost.

This contention, however, understates the significant expansion of the Synthesis in the 1960s to reconsider carefully and debate questions such as those concerning the levels at which selection operates, including genes, individuals, and groups. This broader focus allowed for the exploration of the evolution of social behavior and altruism (topics for which evolution's implications had been clear to Bryan in the 1920s). And soon, the Synthesists' students, in what proved to be transformational work, incorporated advances in molecular biology and genomics (the study of entire genomes), making use of tools that could measure gene products (e.g., protein electrophoresis, a test identifying specific proteins in the blood based on their electrical charges.)[25]

Nonetheless, Gould illustrated his claim of a hardening with the initial response by some Synthesists, especially Mayr,[26] to Motoo Kimura's "neutral theory" hypothesis. Kimura (1924–1994), a pioneering Japanese mathematical geneticist, focused on molecular aspects of evolution. His work led him to question the prevailing view that all genetic changes are driven by natural selection, and he argued that a significant portion of genetic variation is due to neutral mutations. Such alterations would not significantly impact an organism's fitness because they involve a subset of DNA changes that do not affect gene function. Kimura's work introduced the concept of genetic drift at the molecular level, leading to a much deeper awareness of the nature of genetic variation.[27] His thinking had profound implications for understanding evolution at the molecular level, suggesting that not all genetic changes had adaptive consequences. Kimura's work also laid the groundwork for the study of "molecular clocks,"[28] used to estimate the timing of evolutionary events from genetic data.

But Mayr would have nothing of the idea, and according to Gould's eye-witness account, summarily dismissed it at an annual meeting of the Evolution Society in New York, claiming that neutrality was impossible because, as he simply asserted, it was widely known that all substantial change is adaptive. One might question Gould's broad claims in this particular case—perhaps the anecdote says more about Ernst Mayr at that point in his career than it does about any recalcitrant state of the larger Synthesis movement. In any event, although some debate continues around Kimura's neutral theory, it is today widely accepted.[29]

Gould went on further to illustrate the fortification of the Synthesis by examining the way the key but challenging concept of adaptation came to be viewed and studied. He contended that there was a clear shift from hypothesis testing based on adequate data to sloppy presupposition involving "an a priori assumption of near ubiquity." Gould was, to be sure, famous for his wide-ranging critiques of what he called "just so stories," where claims in the literature about adaptation were based on weak, or sometimes nonexistent, evidence.[30] But, in fact, compelling studies of adaptation beyond the peppered moth experiment were beginning to mount.

MORE EMPIRICAL DEMONSTRATIONS

Key research took place—where else?—on the Galápagos Islands that had been so important to Darwin during his time on the *Beagle*. There, the decades-long research of Peter and Rosemary Grant[31] on the beaks of Galápagos finches illustrates the type of evidence that can be obtained in support of natural selection. Their work involved meticulous monitoring of the finches and the documentation of changes in beak size and shape in response to varying environmental conditions.[32] They rigorously measured the beak dimensions of individual finches and tracked changes in these traits over multiple generations.

The key to their groundbreaking findings lay in the Grants' ability to relate these beak changes to fluctuations in environmental conditions. One of the most significant aspects of their research was the close observation of environmental factors such as food characteristics and availability. The islands experienced periods of heavy rain followed by extended droughts, leading to variations in the types of food resources available to the finches. These environmental shifts influenced the availability of different seed sizes and hardness, which directly impacted the finches' foraging.

During periods of drought when smaller, softer seeds became scarce, finches with larger beaks had an advantage because they could crack open and consume tougher, larger seeds. In contrast, finches with smaller beaks struggled to find alternative food sources. This selective pressure led to changes in the finch population, with an increase in the frequency of larger-beaked individuals. Conversely, when heavy rains led to an abundance of smaller, softer seeds, finches with smaller beaks gained a competitive edge as they could exploit these resources more efficiently. This fluctuation in food availability and the subsequent shifts in beak size provided concrete evidence of natural selection occurring rapidly within the finch population. Key to the significance of the results was the compelling evidence that shifts in beak size were inherited.

The Grants' research is consistent with the hypothesis that beak morphology in the medium ground finch was highly responsive to environmental changes, illustrating the principle of adaptation through natural selection. Their work highlighted the dynamic nature of evolution and the critical role of environmental factors in shaping the traits of organisms over remarkably short periods.

An important limitation of this evidence is that it is correlational. In general, field studies involve complex ecological systems with multiple variables at play. Correlations observed in such systems can be influenced by various confounding factors that are not measured or controlled for. For example, in addition to the nature of food and its availability and distribution, factors like competition between species, disease prevalence, or environmental fluctuations could also impact the observed correlations but may not be explicitly considered in the analysis. To establish *causation*, researchers often need to complement observational field studies with controlled experiments.

Other research on Trinidadian guppies (*Poecilia reticulata*) illustrates how combined approaches can work. Predation pressure varies among the streams these fish inhabit and at different points along each stream (because barriers such as waterfalls essentially segregate populations). Differences in spot patterns, colors, and mating strategies have emerged from the divergent selection pressures experienced. Males have smaller, less contrasting, and less variable spots in streams and points inhabited by their major predator, the pike cichlid, than males do in streams that lack them. In locations where there are numerous predators, the males are better camouflaged.

Groundbreaking work on these guppies was conducted by John Endler, whose seminal book, *Natural Selection in the Wild*,[33] bridged an important gap between theoretical models that had been in place since the dawn of the Synthesis and the expanding empirical evidence for natural selection.[34] Endler translocated 200 guppies from a pike cichlids–inhabited stream to an area where the predators were absent. After fifteen generations (about two years), both male spot size and the diversity of color patterns had increased, and the population became more similar in appearance to those inhabiting naturally predator-free streams.

Endler followed up with microevolutionary studies of guppies in large artificial ponds with gravel beds, where some contained coarse gravel and others had fine, multi-colored gravel. He populated the ponds with guppy "mutts" produced by mating for several generations all the naturally occurring spot varieties he had collected in the field. Six months later, in controlled experiments, pike cichlids were introduced into some of the ponds. In others, a less dangerous natural predator, the killifish, was inserted. Control ponds remained predator-free.

Over about ten generations, Endler's findings were striking. In ponds with predators, the number and brightness of spots decreased, and the guppies would have larger spots when the gravel was coarse than when it was fine. Thus, in environments with predators, the guppies evolved patterns that helped them blend with their surroundings. In ponds without their enemies, the number and brightness of spots increased, and guppies had larger spots if in a fine-grained gravel tank and smaller spots if on a coarse-grained gravel setting. These more conspicuous patterns enhanced mating success by making guppies stand out against the background. In a subsequent experiment using the same pond environments but without predators, sexual selection favored male guppies with more distinctive and elaborate patterns, which made them more noticeable to potential mates. For these guppies, there were clear evolutionary consequences of conflicting selection pressures concerning predators and mate attraction.

Clearly, evidence for adaptive traits resulting from natural selection is available and attainable with the proper approach and effort. This evidence was meticulously assembled during the late Synthesis. This period was also a high-water mark for evolution in the courts, although key legal cases like *Epperson* never turned on the empirical validation that was happening contemporaneously.

But an important research area for which that quality of evolutionary evidence has been elusive is human behavior—an area where many have long sought to apply evolutionary ideas, going back to (and before) Darrow's arguments on behalf of Leopold and Loeb and Bryan's passionate criticism. This lack of quality evidence for behavioral adaptations in our species has led to vigorous debates over some claims. Although a lot of the heat and fury over evolutionary interpretations of human behavior centers around the divisive and perpetually controversial topic of genetic determinism[35] and human variation, the ancillary problem of evidence for claims about *adaptation* has also been prominent and has remained contentious over some fifty years. We return to these bitter skirmishes later.

CHAPTER THIRTEEN

A Punctured Synthesis

> What right has a little irresponsible oligarchy of self-styled "intellectuals" to demand control of the schools of the United States?
>
> —William Jennings Bryan, planned closing argument in Trial of John Scopes

ONLY TWO YEARS AFTER *LEMON*, Texas plaintiffs sued for a court order that would block that state's education board from teaching evolution "without teaching the other theories regarding human origin." In a brief opinion, a federal appellate court rejected the case, holding that it would not do as a court something that could not be done by legislatures: "prevent teaching the theory of evolution in public school for religious reasons."[1] But this approach—attempting to condition the teaching of evolution on the teaching of other explanations rooted in creationism—quickly became a prominent strategy for anti-evolutionists moving forward.

THE RETURN TO TENNESSEE

Soon afterwards, and more consequentially, the Tennessee legislature moved to replace the Butler Act with a replacement statute seemingly tailored to comply with *Epperson*. Tennessee's new-model law was relatively unsophisticated. Enacted in

1973, it provided that any biology textbook that introduced a theory relating to the origins of humanity must identify it as only one of competing theories, and textbooks must also balance such theories "including, but not limited to, the Genesis account in the Bible." The law also forbade (but did not define) the teaching of "occult or satanical beliefs of human origin."

It was not on the books for long. Challenges were brought in both federal and state courts. In 1975, a federal appellate court held that, despite surface-level differences in this law from the one applied during *Scopes*, "the purpose of establishing the Biblical version of the creation of man over the Darwinian theory of the evolution of man is as clear in the 1973 statute as it was in the statute of 1925."[2] Relying on both *Epperson* and *Lemon*, the court struck down the bulk of the new law, although it found no need to rule on the "occult or satanic beliefs" provision. That provision did not survive for long after. It was struck down by the Tennessee Supreme Court a few months later, in an opinion making the—perhaps obvious—point that it "would be utterly impossible for the Textbook Commission to determine which religious theories were 'occult' or 'satanical' without seeking to resolve" theological questions.[3]

Anti-evolutionists continued to experiment with new strategies. In 1973, an opponent of evolution tried to challenge, based on standing as a taxpayer, the National Science Foundation's funding for textbooks that contain evolution, arguing that teaching evolution was itself a violation of the separation of church and state. A federal court dismissed his case without issuing an opinion, finding it meritless.[4]

FOSSIL CONTROVERSIES AND GRADUALISM, AGAIN

While the anti-evolution movement was experiencing courtroom failures, a rather public, partial unraveling of the unity within the Modern Synthesis began around this time, with a

1972 paper with the unassuming academic title "Punctuated Equilibria: An Alternative to Phyletic Gradualism."[5] Its authors, paleontologists Niles Eldredge and Stephen Jay Gould, argued that the gradual changes associated with Darwin's concept of natural selection were virtually nonexistent in the fossil record. Instead, what was clear were long periods of steady-state equilibrium characterizing the history of most fossil species. Their proposal built on Ernst Mayr's emphatic focus on geographic speciation, in particular, the notion that new species might often originate not from a main population, but instead from smaller, peripherally isolated sub-populations. Such small populations, separated from the main body of the species and responding to a novel environment or set of conditions, could readily undergo rapid change. Trait changes might occur as a graded continuum in that a complete series of intermediate states from Point A to B occurs, but this would happen so rapidly that the fossil record appears discontinuous.[6]

Gould and Eldredge reasoned that in such cases the fossil record would not likely preserve evidence of this rapid change due to both the suddenness of the events and the limited geographical area involved. However, if these newly evolved species were to subsequently have the opportunity to expand into the primary territory of the parent species, especially if environmental conditions were in their favor, the newcomers might swiftly outcompete the parent population. This would result in (and present to paleontologists as) an abrupt appearance in the local fossil record.

The absence of transitional fossils in such cases had typically been explained by the argument Darwin himself had made, namely, that the fossil record will always and necessarily be incomplete given the rare mechanisms through which fossils are formed. Gould and Eldredge maintained that their model better explained patterns in the fossil record. The paper's challenge to the prevailing gradualist view of natural selection sparked a lively debate that would continue for years to come, attracting significant attention. In retrospect, the debate over Gould's and Eldredge's proposal ought to have been seen as reflecting the dynamic nature of scientific discourse, underscoring the importance of challenging prevailing paradigms, exploring new ideas, and subjecting hypotheses to rigorous scrutiny. But what might at first appear to be a relatively minor modification of traditional evolutionary thought became a major and quite public disagreement.

Once again, the fact that these new ideas differed from Darwin's mattered to outside observers. Darwin, as we have seen, was such a strong advocate for gradualism that any challenge to it was perceived as tantamount to a rejection of Darwinism and, by extension, evolution itself. Not by professional biologists, of course, but likely in the eye of the public, thereby weakening the scientific status of evolution after the transformation of the synthesis.

From an outside perspective, the field could be seen as returning to something resembling its *Scopes*-era internal disharmony. Gould, whose research expertise was speciation in snails, was a superbly talented writer. He commanded a large audience with his regular column *This View of Life* (the title a quote from *Origin*), which ran in *Natural History* magazine from 1974 through 2000, with a total of 300 essays. Gould's prominence—for the public, he *was* the modern face of evolution—made this debate more newsworthy than would otherwise have been the case. This, in turn, gave the general impression that evolution itself was on unstable ground, leading some popular accounts to make direct links between punctuated equilibrium and a new spread of creationism.[7] It surely did not help when the debate took an unusually heated turn when some critics of the punctuated equilibrium proposal described it as "evolution by jerks," which led the more than willingly combative Gould to retort that traditional phyletic gradualism was "evolution by creeps."[8]

Separate from this very public vitriol, the importance of punctuated equilibrium depends upon how common the pattern described by Gould and Eldredge actually is: if it is a rarity, the theory is less consequential and represents a more modest challenge to traditional Darwinian gradualism. Certainly, instances where new taxa appear rapidly in the absence of fossil evidence were recognized in the Modern Synthesis,[9] but much evidence also points to the fact that adaptive change can occur within populations *without* speciation resulting (e.g., the beaks of Galápagos finches). The question of whether changes in traits *routinely* (a rather vague criterion, to be sure) result in the divergence of species into new ones is a core test of the theory of punctuated equilibria, and that assessment is still underway. Many textbooks still tend to stress the "rival ideas" take on this debate, and no decisive winner seems likely soon.

The concept of punctuated equilibria has not in any sense supplanted Darwinian gradualism, as Gould eventually accepted.[10] However, it retained a role in what emerged as a pluralistic account of speciation, of no surprise to modern supporters of the Synthesis. That the debate played out the way it did, under the full glare of the media and public scrutiny, was not inconsequential as a depiction of evolutionary science.

COMPETITION VERSUS SYMBIOSIS

Just prior to the controversies generated by Gould's and Eldredge's punctuated equilibrium paper, another challenge emerged directed at the core conception of evolution as a gradual process driven primarily by competition among individuals. It would take time, however, for its impact to be felt. The challenge originated in the theory of symbiogenesis, or the endosymbiotic origin of eukaryotic cells, emphasizing cooperation over competition as a driving force in evolution. This theory, popularized by Lynn Margulis, proposed that eukaryotic cells (cells with a nucleus and organelles) originated through a process in which different species of bacteria came together in symbiotic relationships, eventually merging into a single, more complex organism.

The idea that complex cells could have evolved through such mergers was not entirely new. The concept of symbiosis as a driving force in evolution was initially explored by the Russian botanist Boris Kozo-Polyansky, who published a book on the topic in 1924.[11] Despite this early work, the idea of symbiogenesis remained largely ignored in the West, where Darwinian competition was the dominant narrative.

Margulis resurrected and significantly expanded upon the concept. Her groundbreaking 1967 paper in the *Journal of Theoretical Biology* proposed that organelles such as mitochondria (the cell's energy producers) and chloroplasts (organelles responsible for photosynthesis) were once free-living bacteria that had entered into symbiotic relationships with larger cells.[12] Over time, these bacteria became fully integrated into their host cells, leading to the evolution of the complex eukaryotic cells that make up all animals, plants, and fungi. Margulis saw Gould's and Eldredge's punctuated equilibrium as a valuable concept for describing the discontinuity in the appearance of new species,[13] and she contended that symbiogenesis was evidence that such discontinuities were real.[14]

Margulis's theory was initially met with widespread skepticism and outright hostility.[15] Her manuscript was rejected by fifteen journals before it was finally published, and she struggled to secure funding for her research. She later recalled one of the more memorable rejections she received: "Your research is crap. Don't ever bother to apply again."[16] Margulis's bold, often combative personality did not endear her to the scientific establishment, but her status as a pioneering woman in a field that remained male-dominated no doubt contributed as well. Unafraid to criticize the prevailing orthodoxy, she once described neo-Darwinism as "a minor 20th-century religious sect within the sprawling religious persuasion of Anglo-Saxon Biology,"[17] (unintentionally echoing the anti-evolutionist legal theory that teaching evolution itself violated the separation of church and state[18]). She was particularly critical of the view that evolution was driven solely by random mutations and natural selection, arguing instead that symbiosis and cooperation were equally, if not more, important in driving evolutionary change.[19]

Margulis's willingness to challenge established ideas earned her a reputation as a scientific heretic. In the scientific community, being labeled a heretic is often the price one pays for pushing the boundaries of conventional thought. Margulis, however, embraced the role, seemingly reveling in it. Her defiance was not just intellectual but deeply personal. She was known for her sharp tongue and unapologetic critiques of her colleagues. She famously dismissed John Maynard Smith, a giant in the field of evolutionary biology, as "codifying an incredible ignorance," and described his work as "reminiscent of phrenology."[20] She believed that the Modern Synthesis had become too focused on mathematical models and statistical abstractions, losing sight of the actual biological processes that drive evolution.[21]

Despite the initial resistance, evidence in support of Margulis's theory accumulated over the years. It was shown, for example, that mitochondria have their own DNA, which is strikingly similar to that of certain bacteria. This, along with other evidence, led to the gradual acceptance of the endosymbiotic theory as scientific orthodoxy. By the time she published her 1981 book, *Symbiosis in Cell Evolution*, Margulis's ideas had gained widespread acceptance. As Ernst Mayr noted, the "evolution of the eukaryotic cells was the single most important event in the history of the organic world."[22]

However, Margulis's broader view of symbiosis as the *primary* driver of evolution remains controversial. She argued that the major leaps in evolution, such as the origin of new species, were not the result of gradual changes through natural selection but rather the result of symbiogenesis, sudden, significant changes brought about by the merging of different organisms.[23] This view put her at odds with mainstream evolutionary biologists, who largely adhered to the neo-Darwinian synthesis. As a result, Margulis often found herself on the fringes of the scientific community despite the eventual acceptance of her endosymbiotic theory. Moreover, some worried that her challenge to a central tenet of natural selection might exacerbate the public's already tenuous understanding of the nuances of evolutionary theory.

Ultimately, Margulis's theory of symbiogenesis not only revolutionized our understanding of the origins of eukaryotic cells but also forced a rethinking of the broader processes of evolution. Her insistence on the importance of cooperation and symbiosis as driving forces in evolution challenged the prevailing view of evolution as a process driven solely by competition. This perspective, although still controversial, has opened up new avenues of research and has led to a more nuanced understanding of the complexity of evolutionary processes. In the end, Margulis's work was important, reminding us that evolution is not just about the survival of the fittest but, at times, also about the survival of the most cooperative.

NATURE VERSUS NURTURE, AGAIN

Meanwhile, the application of evolutionary theory to human behavior was also gaining attention for the bitter infighting that occurred among academics. As we have seen, this debate was not exactly new. Connecting human behavior to evolution was one of the very strategies Darrow had employed to defend Leopold and Loeb, and it had been a popular topic among intellectuals in the early twentieth century. The potential for controversy here had always been immense. Bryan had thus charged in 1925 that "[p]sychologists who build upon the evolutionary hypothesis teach that man is nothing but a bundle of characteristics inherited from brute ancestors."[24] The groundwork for controversy over sociobiology had been laid even at the time of *Scopes*.

In a groundbreaking 1975 book *Sociobiology*,[25] the Harvard entomologist Edward O. Wilson, one of the world's leading ant experts, christened the new field of sociobiology: an extension of population biology and evolutionary theory to social organization and behavior. The field's major premise (quite innocuous sounding in and of itself) was that behavior had been the target of natural selection and could be studied and understood just like other adaptations.

Sociobiology challenged the prevailing "blank slate" notion of behavior, that human actions and cognition are governed by knowledge acquired only through experience. Its application to humans sparked a significant controversy that continued well into the next decade. The "sociobiology debate" (some say "wars") had Wilson and his growing number of allies facing accusations of racism, misogyny, and support for eugenics.[26] Although animal sociobiology was largely accepted, critics feared that biological explanations for human behavior might inadvertently justify social inequalities or diminish human agency.

In a piece on sociobiology entitled "Genes über Alles," *Time* magazine noted that although Wilson had yet to "achieve the stature of a Darwin," his movement had sent "the same kind of shock waves through the academic community."[27] *Time* went on to describe how sociobiology's position was that "some—and perhaps much—of human behavior is genetically determined," although Wilson's actual stated contention at the time was that at most 10% to 15% of human behavior was genetically based.

Much like Darwin had done in *Origin*, in *Sociobiology*'s almost seven hundred pages, Wilson assembled a vast amount of information about ecology, life history patterns, ethology, sociality, mating, and parental care across diverse taxa. He integrated it all within the Modern Synthesis, including the recently expanded theoretical reaches of evolutionary biology, such as kin selection, reciprocal altruism, and parental investment. Wilson also provocatively challenged the social sciences. Even more audaciously, to his critics at least, Wilson proclaimed:

> Scientists and humanists should consider together the possibility that the time has come for ethics to be removed temporarily from the hands of the philosophers and biologicized. The subject at present consists of several oddly disjunct conceptualizations.[28]

Thus challenged, it is no surprise that the offended academics did not take kindly to outsiders—an entomologist, of all people—informing them that their fields and views needed desperate help.

Wilson had not exactly issued a declaration of war, but the response was as swift and intense as if he had. One of the first to retaliate was cultural anthropologist Marshall Sahlins, who quickly fired off a slim book expressing concern that sociobiology had "an intrinsic ideological dimension . . . a profound historical relation to western competitive capitalism."[29] Sahlins focused his attack on a central theory showcased in the nascent sociobiological field, kin selection, an idea that Darwin had vaguely alluded to and had been hinted at by Haldane and Fisher, but was more formally explained and modeled later. In 1964, Biologist W. D. Hamilton realized that because relatives shared genes by virtue of their common ancestors, natural selection could promote nepotistic altruism through this additional route of gene transmission. Under specific conditions that Hamilton specified, an individual that helped their kin in turn helped their genes. (Here, there are echoes of Bryan's observation, at *Scopes*, that "reciprocity is a calculating selfishness."[30])

Sahlins countered that such a kin calculus could not possibly account for human kinship classifications or patterns of social behavior because, he claimed, no system of human kinship is organized in accord with the genetic coefficients of the relationship. He argued that established human conceptions of kinship can be so far from biology as to exclude all but a small fraction of a person's genealogical connections from the category of "close kin" while, at the same time, including in that category, as sharing common blood, very distantly related people or even complete strangers. Largely based on just these observations, Sahlins summarily dismissed the emerging sociobiological enterprise in its entirety.

Reviewing Sahlins's book in *Science*, the Synthesist George Gaylord Simpson concluded:

> Much of the critical discussion of sociobiology, including this book, has been another form of the nature-nurture debate, a discussion that has proved futile and indeed meaningless because that is not a legitimate either-or question. Man is not born a tabula rasa, nor is he born a programmed automaton. When the argument approaches that extreme polarization, it is sensible to say, "A plague o' both your houses."[31]

Gould was also a staunch critic of the new sociobiological approach. Along with his colleague at Harvard, evolutionary geneticist Richard Lewontin, he viewed sociobiology as intrinsically naive and imminently dangerous when applied to human behavior, with some examples being the extreme and inflammatory claims (not made by Wilson himself, but rather emerging from the developing field of evolutionary psychology that would later be inspired by the sociobiology movement) that even rape and child abuse might be adaptive traits. Gould and Lewontin construed such claims as a repackaging of the sort of genetic determinism that underpinned eugenics. Lewontin described his view of sociobiology bluntly:

> This is fundamentally a very conservative worldview, which serves the very important function of saying that there is no sense in rocking the boat—we are what our genes make us—and I think that's bullshit.

It is ironic that for Darwin in 1859, evolution was clearly the foe of conservatives, but in 1975, it was cast as an ally. But as heated discussions between Darrow and Bryan demonstrate, evolution has long been a source of controversy not just because of its implications for human *origins* but also because of evolutionary claims about human *nature*.

Over the course of the 1970s, things heated up even more. Members of the American Anthropological Association went so far as to consider a formal motion to censure the field of sociobiology and cancel two symposia on the topic at their 1976 meeting. The resolution also denounced sociobiology's "pernicious influences on the young, through its use in school texts,"[32] words that could have easily come from the mouths of Bryan and his allies. The debate continued for an hour, but the resolution was defeated, in part because to some members, aware of the obvious historical comparison, it was too reminiscent of the *Scopes* trial.

The acrimony continued and even intensified when, during a session on sociobiology at the 1978 annual meeting of the American Association for the Advancement of Science, Wilson was confronted by some two dozen protesters, subjected to accusatory chants, and drenched with water from a pitcher.[33] Additionally, the International Committee Against Racism[34] leveled accusations against Wilson, alleging that he promoted racist ideas and advocated for genetic determinism.

Nonetheless, despite massive scrutiny and repeated criticism over the ensuing decades,[35] sociobiology not only survived but expanded, although few researchers actually refer to themselves as sociobiologists today, no doubt simply to avoid residual stigma from the most vitriolic years. Methodological advancements were pivotal in sociobiology's ultimate success. The initial heavy reliance on theory and broad comparative analyses in animals expanded to include genetic, neurobiological, and computational methods. In human sociobiology, interdisciplinary research integrating genetics, psychology, and anthropology has provided far more nuanced insights into the interplay between biology and culture. Genomics and molecular biology have significantly advanced research on the genetic basis of social behavior. By identifying specific genes and neural pathways associated with social behaviors, researchers have gained deeper insights into the mechanisms driving these behaviors. This kind of evidence has allowed for far more sophisticated research into adaptation, as well.

Sociobiological explanations spread quickly to other fields as well. By 1980, University of Chicago law professor Richard Epstein, although critical of evolutionary explanations for some legal doctrines, conceded that the "gene-environment interactions that drive natural selection" had likely selected for individuals with "tastes for legal rules," such as prohibitions against the use of force outside of self-defense.[36] Two years later, William Rodgers Jr. of the University of Washington would cite Wilson's *Sociobiology* to push back against what he perceived as legal theorists' overreliance on the work of John Rawls.[37] Citing Wilson's work, Rodgers argued that Rawls's analysis of "a state of nature, and a calculated social compact stemming from it," was "utterly at odds with what paleoanthropology knows of the evolution of the human species." Like Epstein, Rodgers looked to evolution to explain why human nature developed to be compatible with legal rules.

The catastrophic impact on society feared by many critics in the mid-1970s did not transpire—some might say, because of the intense backlash sociobiology received and the field's increased awareness, sensitivity, and response to criticisms that highlighted the initial debate. In the eyes of some disciples, sociobiology triumphed,[38] even producing academical progeny, including the field of evolutionary psychology. Wilson's last chapter in *Sociobiology*—the one focused on humans and prompting such a harsh reaction—was the opening salvo in this particularly long, drawn-out battle over human nature, which was far from over.

The Modern Synthesis solidified Darwinian natural selection as the cornerstone of evolutionary biology. But the ensuing years marked a return to internecine squabbles and controversy. These internal debates, natural and crucial for the advancement of any field of science, had far-reaching consequences beyond the scientific community, notably affecting the public perception of evolution and drawing attention from evolution's omnipresent opposition, the creationists. Dissent within evolutionary biology has historically revolved around the mechanisms of evolution, rates of evolutionary change, and the relative importance of different evolutionary processes. For a larger audience, the major contentious issue has become the hereditary basis of human nature, which remains a keenly debated topic.

These dissensions, although integral to scientific advancement, can be misinterpreted by the public and, importantly, manipulated and misrepresented by those advocating for particular world views, as uncertainty or weakness in the foundational principles of evolution. The public's understanding of science largely comes from media representations of it, which many times do not accurately reflect the nature of scientific debate. Media reports often sensationalize disagreements, leading to a perception that the entire concept of evolution is in dispute. These phenomena would play out in another wave of *Scopes*-style anti-evolution cases brought in the 1980s and heard in the aftermath of the Synthesis's partial unraveling.

CHAPTER FOURTEEN

Crusades Begin

> Mr. Bryan . . . is a great leader . . . [W]hen any great leader goes out of his field and speaks as an authority on other subjects, his doctrines are quite likely to be far more dangerous.
>
> —Argument of Dudley Field Malone, Trial of John Scopes, Day 5

IN 1980, TWO ANTI-EVOLUTION organizations sued the Smithsonian Museum of Natural History, attempting to block display of an exhibit that referenced and explained evolution. Federal courts rejected that challenge like they had rejected anti-evolution lawsuits throughout the 1970s.[1] But another wave of attacks was only beginning. In no way was it mitigated by the large body of sophisticated evidence that now supported evolution. On the contrary, opponents would at times cite internal controversies, such as the renewed debate over gradualism, in attempts to discredit the field as a whole.

ARKANSAS, AGAIN

In 1968, Arkansas attorneys had barely bothered to defend the state's *Scopes*-era law before the Supreme Court, drawing the ire of Justice Black. But in 1981, state leaders enacted a new-model anti-evolution statute, replacing the one that had fallen in *Epperson*.[2] This new act reflected years of further strategic refinement by the anti-evolution movement, which had learned from the failed recent attempt in Tennessee. Arkansas's new law would not explicitly mandate the teaching of Genesis (although it mentioned a "worldwide flood"), nor would it hyperbolically refer to "satanical" theories of human origin.

Instead, it mandated balance between two expressly defined theories. The law defined "Creation science" using benchmarks that overlapped almost entirely with fundamentalist religious doctrine, including sudden creation from nothing, separate ancestry for humans and apes, recent inception of the Earth, and geological features caused by catastrophes including a worldwide flood. Evolution was defined as a separate science, and the law required "balanced" teaching of both.

Although this measure was more sophisticated than Tennessee's earlier effort, it fared no better. A group of plaintiffs organized by the ACLU challenged the law in federal court in 1981. Judge William Overton conducted a ten-day trial, hearing extensively from expert witnesses on evolutionary biology and creation science. One of these witnesses was Stephen Jay Gould, whose punctuated equilibrium theory—that rapid evolutionary bursts were followed by long periods of stasis—and the resulting controversy within the field had been observed and noted by creationists who supported Arkansas's law.

As Gould prepared to testify, in June of 1981, he fittingly traveled to Dayton, visiting Robinson's Drug Store where Scopes's journey had begun.[3] The store remained one of Dayton's social hubs, although it had moved, in 1928, from its original location to be nearer to the now-legendary Rhea County Courthouse. Sonny Robinson, five years old in 1925, had taken over the family business. As Gould took photographs, Robinson came to realize that the visitor was no ordinary tourist but a famous professor of evolution from Harvard. He made calls, much as Rappleyea once had, and assembled a group of local notables for photographs. It was brutally hot—a humid 97°F—just as it had been at the *Scopes* trial, but Gould played along.

Gould's public appearances and writings at the time, including his trip to Dayton and an editorial published in the *New York Times*, reflected his conviction that the *McClean* case had broad implications for science education and the public's understanding of evolution.[4] His testimony at trial demonstrated the remarkable breadth of his knowledge and scholarship; he was a well-known paleontologist and theorizer, a celebrated popularizer of natural history and evolution, and a historian of science. At the trial, Gould placed the long-running debate between evolution and fundamentalists in a historical context. He stressed that his views on punctuated equilibrium were being distorted and misused, noting strong historical parallels to *Scopes*. In 1925, Bryan had similarly distorted William Bateson's public speeches to suggest that scientific belief in evolution was tenuous.

After hearing testimony, in a voluminous opinion, Overton methodically applied each prong of the *Lemon* test, which required him to determine if Arkansas's anti-evolution law had a secular purpose, if it advanced religion, and if it would entangle the state with religion. These were not straightforward questions because Arkansas's new law had been carefully drafted to avoid the explicit Biblical references that had undermined prior anti-evolution laws. Moreover, although Fortas did focus on the impermissible purpose of the challenged law in *Epperson*, judges have long recognized the difficulty of identifying the "purpose" of any law—particularly when asked to strike down a law because its purpose was allegedly improper.

To be sure, legislators routinely make statements and give speeches about why they support bills that are under consideration. Indeed, the sponsor of Arkansas's new-model anti-evolution law had frankly and explicitly connected the bill to his own religious beliefs.[5] Nevertheless (in the words of Chief

Justice Earl Warren), what "motivates one legislator to make a speech about a statute is not necessarily what motivates scores of others to enact it."[6] And overturning a law because of speeches made by its supporters is a problematic undertaking because the same law might be enacted later with fewer (or more circumspect) speeches made in its support, thus causing overall confusion. Thus, although religious purpose was clearly relevant under governing legal tests, a few religiously tinged speeches from the law's drafters or sponsors might not be enough to doom it.

Overton thus faced a more nuanced and complicated problem than Fortas had dealt with, but he took his job seriously and did it thoroughly. In *McLean v. Arkansas*,[7] he produced the most comprehensive judicial examination of the evolution issue to date. (Interestingly, it is one of the only written judicial decisions dealing with evolution disputes that does not explicitly reference the *Scopes* trial.) First, Overton charted the history of fundamentalist opposition to evolution, how that opposition gave rise to the organized "scientific creationist" movement, and how that movement influenced the drafting and enactment of Arkansas's new law. Here, he benefited from Gould's testimony and expertise in both evolutionary theory and the history of science. Overton then examined the structure of Arkansas's law, finding that its definition of creation science was inescapably religious, notwithstanding the lack of any explicit mention of the Bible. He described the statute's division between evolution and creation science as "a contrived dualism" with "no scientific factual basis or legitimate educational purpose."

Overton went on to craft a judicial test for identifying a legitimate science, using characteristics such as falsifiable claims. He explained that the challenged law's definition of "creation science" did not, in fact, refer to actual science because it "is not explanatory by reference to natural law, is not testable and is not falsifiable." The sum total of these conclusions established that Arkansas's law failed all three parts of the *Lemon* test. Despite its sophistication, in the end, Arkansas's law fared no better than Tennessee's had done.

THE SUPREME COURT, AGAIN

Louisiana was next to enter the fray. In 1981, that state enacted perhaps the definitive version of a new-model anti-evolution statute, setting off a six-year legal battle that would ultimately bring the issue back before the nation's highest court. Louisiana's law did not require schools to provide any instruction on humanity's origins, but any school that did so was obligated to give "balanced treatment" to evolution and creation science. The law also banned so-called "discrimination" against any teacher who chose to "teach scientific data which points to creationism" (reminiscent of the original and recurring problems with an academic-freedom view of the *Scopes* controversy). It required the governor to select seven creation scientists to assist schools in developing a creation-science curriculum. Not only did the law avoid Biblical references, but it also avoided Arkansas's mistake of explicitly referencing Biblical events like the world flood.

Everyone knew this was headed for litigation. Two competing lawsuits were quickly filed. The bill's sponsors sought an affirmative ruling that the law was constitutional in a federal court in Baton Rouge, while plaintiff organizations supported by scientists challenged the law's constitutionality in a federal court in New Orleans. Initially, both courts dodged the most controversial issues presented by their respective cases. The Baton Rouge case was dismissed on technical grounds, and the court in New Orleans initially struck down the statute for a narrow reason, holding only that it violated structural provisions of the state constitution.[8] Neither decision delved into a scientific debate as Judge Overton had.

The New Orleans decision was referred to the Louisiana Supreme Court, which had final authority to interpret the state's constitution. It held that the law did not violate the structural provisions identified by the New Orleans court, deciding that the legislature could validly regulate "courses of study" in public schools, so long as those regulations were constitutional (an issue the court did not resolve).[9] Justice John Allen Dixon grumbled that his judicial colleagues were dodging the hard issue at the core of the case, arguing that creation science was clearly a religious doctrine, not a "course of study."

The case returned to New Orleans. The state defendants now sought a trial to determine what they considered a disputed issue: the definition of "science." But Judge Adrian Duplantier, a former state senator appointed to the federal bench by Jimmy Carter, ruled that a full trial was unnecessary and the law was invalid. He did not follow Overton's lead in attempting to define science in a legal opinion. Instead, he wrote: "Whatever 'science' may be, 'creation,' as the term is used in the statute, involves religion." Thus the law was an impermissible religious establishment.[10]

Duplantier's ruling was just the beginning, and Louisiana was determined to pursue all available options. First, the state appealed to a federal appellate court, where a panel of three judges upheld Duplantier's decision. That decision was written by E. Grady Jolly, who had been elevated to the federal bench by Ronald Reagan and was a frequent critic of evolution.[11] But Jolly recognized that Louisiana's law had not been written on a clean slate. No matter how carefully the law had been drafted to avoid explicit references to religion, it was part of a deep historical legacy that was inherently religious. In Jolly's words, the case arose

> against a historical background that cannot be denied or ignored. Since the two aged warriors, Clarence Darrow

> and William Jennings Bryan, put Dayton, Tennessee, on the map of religious history . . . courts have occasionally been involved in the controversy over public school instruction concerning the origin of man. With the igniting of fundamentalist fires in the early part of this century, 'anti-evolution' sentiment, such as that in *Scopes*, emerged as a significant force in our society. . . . this scheme of the statute, focusing on the religious *bete noire* of evolution, as it does, demonstrates the religious purpose of the statute. Indeed, the Act continues the battle William Jennings Bryan carried to his grave.

It was thus, in part, the law's relationship to the legacy of *Scopes* that demonstrated religious purpose and led to the panel's conclusion that it failed the *Lemon* test.[12]

The state defendants were undaunted. They next sought a rare form of relief by inviting all fifteen active judges of Jolly's circuit to review his panel's decision. At this stage, the law's defenders nearly won; a rehearing was denied by a narrow and contentious vote of eight to seven. Judge Thomas Gee, a Nixon appointee, wrote a dissent on behalf of the seven judges who sided with the state. Interestingly, Gee attempted to reverse the *Scopes* legacy, strangely implying that Darrow would have been on the side of Louisiana. Gee wrote, hyperbolically, that the

> *Scopes* court upheld William Jennings Bryan's view that states could constitutionally forbid teaching the scientific evidence for the theory of evolution, rejecting that of Clarence Darrow that truth was truth and could always be taught—whether it favored religion or not. By requiring that the whole truth be taught, Louisiana aligned itself with Darrow; striking down that requirement, the panel holding aligns us with Bryan. . . . It comes as news to me . . . that the Constitution forbids a state to require the teaching of truth.

In a biting response, Judge Jolly sarcastically apologized to the dissenters for failing to align with "their commitment to the search for eternal truth through state edicts."[13]

The law's defenders had one last hope: they sought review by the Supreme Court, which agreed to decide its first evolution case since *Epperson*. Just as in the *Scopes* trial, an assemblage of noted scientists participated through a friend-of-the-court "Brandeis brief," arguing against the validity of creation science. Seventy-two Nobel laureates and other prominent research organizations argued that the state legislature had usurped and supplanted an academic determination of what science is—and isn't.[14]

These were not merely evolutionary biologists defending their turf; their large number included scientists like chemist Glenn Seaborg (whose research on trans-uranium elements led to the discovery of plutonium) and physicist Emilio Segré (who had demonstrated the existence of the antiproton). It also included Francis Crick and James Watson, who famously discovered the molecular structure of DNA.[15] Murray Gell-Mann, whose work on the classification of elementary particles earned a Nobel prize in 1969, organized the recruitment of fellow laureates.

Gell-Mann saw in the anti-evolution movement a much broader danger to science as a whole, in part because the Synthesis had been built upon insights and evidence gathered from a range of disciplines, and evolution's opponents had accordingly expanded their list of targets to include other fields. For example, in Gell-Mann's words, "fundamental and well-established principles of nuclear physics are challenged, for no sound reason, when 'creation scientists' attack the validity of the radioactive clocks that provide the most reliable methods used to date the earth."[16]

Much as Winterton Curtis had during the 1920s, Gell-Mann embarked on a prodigious letter-writing campaign, soliciting support from fellow scientists against, in his words "an attempt to misrepresent science for the sake of promoting fundamentalist religion."[17] The brief took on new importance after oral argument: both Gell-Mann and Gould (who was keenly involved in this case as well) deemed the ACLU's legal argument at the High Court tepid and unimpressive.[18] Gould wrote to Gell-Mann, "Our oral argument was so bad that our only hope now resides in the [written] briefs."[19]

But the scientists' brief was a good one, a classic of the "Brandeis brief" genre. Citing the work of major Synthesists, including Mayr and Dobzhansky, it established with clarity that "evolutionary history of organisms has been as extensively tested and as thoroughly corroborated as any biological concept," making the state's presentation of evolution as a less established theory "scientifically indefensible." And more specifically, the brief took on the anti-evolutionists' misrepresentation of the punctuated equilibrium controversy.

In defense of "creation science," the state's defenders had argued that it could validly present a "nonreligious" theory of "abrupt appearance in complex form" as an alternative to evolution, and they had misrepresented the punctuated equilibrium hypothesis as supporting "abrupt" origins. The scientists' brief addressed this nicely:

> [W]hile Darwin thought that the evolutionary change that produces new species occurred over millions of years in large populations, many modern scientists hypothesize that such change takes place over only a few hundreds or thousands of years in limited populations. Those favoring a more rapid pace of evolutionary change are careful to emphasize that their disagreement concerns the proper elaboration of the evolutionary theory, not a choice between creation and evolution. Appellants and their affiants[, meaning, the state and its witnesses,] cite the works

> of various participants in this debate to support their view that "abrupt appearance" is a valid scientific theory. Yet appellants never clarify what they mean by "abrupt." The scholars who use "abrupt" or "sudden" in this debate mean hundreds or thousands of years. If by "abrupt" appellants mean "instantaneous," then they are quoting the authorities out of context. If instead they mean "over only a few hundreds or thousands of years," then "abrupt appearance" is not an alternative to evolution but a part of it.[20]

Justice William Brennan Jr. wrote the Supreme Court's majority opinion. One of the most influential justices to ever sit on the Court, Brennan had been a de facto leader of the Court's liberal wing during its heyday in the 1960s and early 1970s, when it had reshaped criminal procedure, freedom of expression, reproductive rights, and other constitutional touchstones. A liberal of the modern style, his jurisprudence embodied a civil libertarian approach, viewing popular majorities with more skepticism than progressives had during the *Lochner* and Roosevelt eras. Nearly twenty years before, as a younger Justice, Brennan had joined Fortas' majority opinion in *Epperson*. He would reach the same result here.[21]

In *Edwards v. Aguillard*, Brennan wrote that Louisiana's new-model anti-evolution law was unconstitutional.[22] He devoted relatively little space to the scientific issues raised by the case and did not rely on the submission of the Nobel laureates. Instead, his opinion highlighted a different concern: In a series of opinions dating back to the early 1960s, Brennan had long argued that the separation of church and state applied with extra force in public schools, where children are impressionable and attendance is mandatory.[23] Applying the *Lemon* test in light of this additional consideration, Brennan concluded that Louisiana's law "advances a religious doctrine by requiring either the banishment of the theory of evolution from public school classrooms or the presentation of a religious viewpoint that rejects evolution in its entirety." Anti-evolution laws had once again lost.

That said, although Brennan might have seemed like the model judge to reinforce Fortas's *Epperson* opinion, and the evolutionary field was even further developed than it had been in 1968, *Edwards* was a far weaker win for the scientists' side. Brennan closed his opinion with a defensive caveat: "teaching a variety of scientific theories about the origins of humankind to schoolchildren might be validly done with the clear secular intent of enhancing the effectiveness of science instruction."[24] Perhaps this caveat was needed to win over the votes of other justices, who together made up a far more conservative Court than the one Fortas had written for. Nevertheless, the Nobel laureate brief had clearly explained that evolution, as a general concept, was at least as validated as any other biological theory; accordingly, the introduction of other alternative theories could not enhance science instruction. Some have criticized Brennan's opinion as implying that the anti-evolutionist movement might *still* prevail if only they found better (meaning, less obviously religious) theories to rely on.[25] Thus, the battles continued.

ACADEMIC FREEDOM, AGAIN

Yet another series of lawsuits, billed in the media as reruns of the *Scopes* trial,[26] arose in the early 1990s. But roles would now be reversed, with anti-evolutionist teachers casting themselves in the role of John Scopes. These cases would aptly demonstrate why one of the ACLU's initial theories of the *Scopes* trial—that of a teacher's freedom of speech within the classroom—was never a perfect fit for the underlying dispute.

Ray Webster, a social studies teacher in New Lenox, Illinois, became involved in a dispute with his school board in 1987. Webster "taught nonevolutionary theories of creation to rebut a statement in the social studies textbook indicating that the world is over four billion years old." After the board mandated that Webster teach from the established curriculum, which included evolution, the dispute ended up in federal court. The Seventh Circuit recognized that school boards did not have completely unfettered control over teachers' speech, but in this case, the board "had the authority and the responsibility to ensure that Mr. Webster did not stray from the established curriculum."[27]

Shortly thereafter, Los Angeles-area high school biology teacher John Peloza launched a more robust challenge to his school's curriculum. Peloza alleged that "evolutionism" is itself a religion rather than a scientific fact, and that his own religious exercise and free speech rights were violated by the school's requirement that he teach it. The response by school authorities (which contained its own striking inaccuracies) was that the "concept of evolution as taught in the high school classes today is not the simple, traditional Darwinism theory of man's evolution from lower life forms, but encompasses changes in animal life, plants, geologic and astronomic processes."[28] (The board's implication that Darwin's ideas did not encompass "changes in animal life" or discuss geologic processes is, of course, bizarre.)

Peloza's complaint was dismissed by a federal trial court, partially resurrected by an appellate panel, and then dismissed again by the Ninth Circuit, sitting en banc. The Ninth Circuit's opinion echoed the reasoning of Justice Chambliss, in his concurrence to the *Scopes* appeal. It distinguished evolution as "a biological concept: higher life forms evolve from lower ones," and explained that this "has nothing to do with whether or not there is a divine Creator." "Only if we define 'evolution' and 'evolutionism' as does Peloza as a concept that embraces the belief that the universe came into existence without a Creator might he make out a claim." Peloza lost, but the court held that his complaint was not so frivolous as to justify an award of attorneys' fees and costs to the defending school district.[29]

Taken together, the alternative approaches adopted by Webster and Peloza did not directly result in more success for opponents of evolution than the new anti-evolution laws did. These cases did, however, turn a traditional narrative from *Scopes* and *Inherit the Wind* on its head. The teachers invoking the First Amendment and the freedom to speak in the classroom were now opponents of evolution, not advocates for it. These arguments were foreseeable, and demonstrate why the free-speech arguments advanced in *Scopes* and *Epperson* were never as good of a fit for the case as arguments rooted in the separation of church and state.

DAYTON, AGAIN

One of many cases billed as "Scopes II" arose a few years later in Hawkins County, Tennessee, located northeast of Dayton. There, a group of parents objected to a series of basic reading textbooks, designed for elementary and middle school students, that incorporated reading passages from other disciplines. The challenged books were hardly strident in their presentation of evolutionary theory—in fact, the books themselves contained a disclaimer repeating the canard that evolution is "a theory, not a proven scientific fact"—but parents still challenged them for including content they deemed religiously objectionable. These supposedly objectionable passages included ones addressing evolution as well as "magic" and even "biographical material about women who have been recognized for achievements outside their homes."

Constrained by binding precedents such as *Epperson* and *Aguillard*, these parents did not seek to ban the books from the classroom, but they did win a trial court judgment ordering their children to be excused from classes where the offending passages would be taught, along with an eyebrow-raising award of more than $50,000 in monetary damages for purported violations of their constitutional right to religious liberty. A federal appellate court reversed that money judgment in 1987, clarifying that mere exposure to curricular ideas that a student or parent disagrees with does not violate religious rights or give rise to monetary damages.[30] This was neither the first nor last battle over evolution in Tennessee courts.

By the turn of the new century, anti-evolutionists were carrying a long losing streak. But back in Dayton, one bastion for their side held strong. Bryan College was a small Christian school established in honor of the Great Commoner a few years after his death nearby. Its motto fittingly remains "Christ above all." Showing good humor and an awareness of history, its president had come to Robinson's Drug Store when Gould visited, posing for *Scopes*-retrospective photos with the renowned Harvard paleontologist.

By this time, Dayton lay about a thirty-minute drive southwest of the Watts Bar nuclear plant. The facility was owned and operated by the Tennessee Valley Authority (TVA), the governmental organ once championed by eccentric *Scopes* attorney John Randolph Neal. With the town nestled between a religious college dedicated to Bryan and a TVA nuclear plant, even the landmarks seemed to stake out conflicting positions in a continuing duel between tradition and modernism. In between, Robinson's Drug Store had closed in 1983, the year after Gould's visit. And the abandoned Dayton Coal & Iron ovens lay buried beneath a sea of kudzu,[31] an invasive species of ivy whose uncheckable rampage across the American South (where it has easily outcompeted native species) provides a textbook lesson in competition and natural selection.

In the early 2000s, it came to light that, for many years, the same Rhea County public school system that had once employed John Scopes maintained a program called the Bible Education Ministry (BEM). School officials claimed the program was optional, but children were not told so, and in practice, no child had ever opted out. BEM was operated by Bryan College, and its programs were taught by volunteers from Bryan's student body. Through BEM, public school students were thus taught from a lesson plan that included points like the following:

> Turn out the lights and tell the children [to] close their eyes. Explain about the darkness. This is all there was before God created the world. Then God made the sun, moon and stars. Turn on a flashlight to show them how light came into the world. . . . Stress that only God could make all this. . . . Teach kids about the creation of man & the fall of man. . . . Show the kids a bag of dirt and ask them what can be made from it. . . . We want to teach them about how on the 5th day God made the birds and the fish.

Having outsourced the lesson planning and teaching to external volunteers, the school district's attorneys argued that there had been no constitutional violation by the government itself.[32]

Two parents of Rhea County schoolchildren sued. A federal district court, dryly noting that Rhea County was "no stranger to religious controversy in its public schools," held that the school's actions violated *Lemon*. The third part of the *Lemon* test, which prohibits actions that excessively entangle the government with religion, was especially implicated. As Judge Robert Allen Edgar explained, the "wholesale delegation of the administration of that program to Bryan College, a decidedly religious institution, by itself results in an impermissible entanglement of government and religion."[33]

Judge Edgar was not a strident liberal along the lines of Abe Fortas or William Brennan. He had been elevated to the federal bench by Ronald Reagan (as had been E. Grady Jolly). Yet Edgar's decision demonstrates how settled these legal issues seemed to be after so many precedents decided against the anti-evolutionists. He explained that this was "not a close case," and he gently chastised the government attorneys for arguing—much as Steward, Hicks, and the rest once had—that "Rhea

County is a place where they respect the Bible." As Edgar explained, although it was likely true that most residents supported the school board's position, the "Constitution protects each one of us, including those who may not have the same religious views as the School Board." Evolution had come back to Dayton, but US law was greatly changed.

The case illustrated another, darker, change as well. Virtually all observers at *Scopes*—even the cynical and sneering Menken—had conceded that the townspeople welcomed visitors with hospitality regardless of viewpoint. There had been initial safety concerns, and a few extra police had been brought in to provide security. Nonetheless, the many reports from the *Scopes* trial consistently emphasized the geniality of the people of Dayton in 1925. John Scopes had even shared pleasant conversations with his prosecutors, including a chat with William Jennings Bryan about their shared hometown in Illinois.

Cultural strife seemed harsher in 2002. The parents who challenged the Bryan College ministry program convinced the court to let them proceed under the pseudonyms John Doe and Mary Roe for their own protection. A letter to the editor in a local paper demonstrated why. Someone had written an open letter to the anonymous parents, accusing them of being "cowards because you won't give us your name. . . . I would love to come face to face with you because yes I would tell you what I thought of you and I would let my sons tell you too. You have hurt my sons and I will not let no one hurt one of my children [sic]." The local principal told a reporter that, had he known who the parents were, he would have asked them, "Do you want to cause your family trouble? . . . Attack religion and crusades begin."[34] Such tensions—manifested in the threat of violence and the protection of identities—highlight the intensity of the cultural battles surrounding evolution and set the stage for the broader legal and political shifts that followed.

CHAPTER FIFTEEN

Backlash and Unraveling

> The defenders of Mr. Scopes . . . committed the mistake which modernists and liberals of all kinds so often commit of fighting a defensive battle on a terrain which their opponents had selected.
>
> —"The Baiting of Judge Raulston," *New Republic*, July 29, 1925.

MOMENTS OF CONSTITUTIONAL CONSOLIDATION often trigger a reaction, initially in the form of political backlash.[1] In time, backlash-driven political debates influence jurisprudence as well, leading to shifts in the law. The *Lochner* decision, we have seen, appeared to constitutionalize laissez-faire economic philosophy in the early twentieth century. However, contrarian arguments by Holmes and others who believed in jurisprudence more deferential to the democratic processes gained influence in time. *Lochner* is now considered anathema.

Just as *Lochner* fell, the midcentury legal consensus that followed it would fall in turn. Two years after the momentous *Lemon* decision, the Supreme Court decided one of the most famous cases in its entire history. When *Roe v. Wade* constitutionalized the right to an abortion in 1973, the political backlash was fast and fierce. A conservative movement coalesced to challenge midcentury legal dogma and oppose a liberal

Supreme Court that had produced outcomes that Clarence Darrow, Arthur Garfield Hays, and the *Scopes*-era ACLU could have only dreamed of: desegregating the country, designing rigorous procedural protections for accused criminals, strictly enforcing the separation of church and state, deferring to the scientific expertise of administrative agencies, and establishing strong safeguards for reproductive freedom.

The modern conservative legal movement's roots are complex, but it arguably began with a group of attorneys who were discontented with the aggressive liberalism of the Warren Court, including unconfirmed Supreme Court nominee Robert Bork and Reagan-Administration Attorney General Edwin Meese III.[2] In one of the (inexplicably) few works in the scholarly literature comparing the origins and impacts of the anti-evolution and anti-abortion movements, Joelle Anne Moreno explained that "these dual social and public policy movements" shared "contemporaneous trajectories, overlapping constituencies, and near-identical legal strategies."[3] Anti-abortion activists' gradual but steady impact in lending support to a budding new conservative legal movement was transformative, with profound effects on, among other things, evolution law and the legacy of *Scopes*. Early proponents like Bork and Meese championed a constitutional philosophy known as originalism, initially arguing that constitutional provisions must be interpreted according to the original intent of those who drafted them.[4]

Over time, theories of originalism adapted and evolved. The focus on the intentions of a few well-known historical figures was largely abandoned—at least in theory.[5] Most modern originalist thinkers prefer looking to "original public meaning," searching for the public's understanding of words used in the constitutional text at the time when particular constitutional provisions were ratified. Specific subschools of originalism, and nuanced techniques for deriving original meaning, have proliferated. All share at least one central tenant: the meaning of constitutional provisions is "fixed" at the time of ratification, without room for the interpretation of those provisions to shift merely because society has subsequently evolved or scientific discoveries have expanded the bounds of human understanding.[6]

A related intellectual development was the ascendancy of textualism, an interpretive methodology assigning near-total primacy to the plain meaning of the words of a governing document. Textualists may turn to various canons to interpret the meaning of ambiguous phrases,[7] but they broadly deprioritize any independent judicial assessment of the broader purpose a legal document was designed to serve.[8]

The rise of textualism and originalism was, at least initially, very much a conservative project. But advocates of these jurisprudential philosophies deploy arguments that would have resonated with the progressives of Bryan's age. In matters of statutory interpretation, textualists purport to act as faithful agents of the legislature by basing their rulings on the direct and explicit authorizations of the people's representatives.[9] In matters of constitutional interpretation, a stricter focus on the words of ratified text was supposed to reduce the free-wheeling pronouncements of an unelected judiciary, presumably restoring power to the elected branches. A critic of looser approaches to constitutional construction asked a question that might have come from the mouths of Bryan or La Follette: "What secret knowledge, one must wonder, is breathed into lawyers when they become Justices . . . that enables them to discern that a practice which the text of the Constitution does not clearly proscribe . . . is in fact unconstitutional?"[10]

That critic was Antonin Scalia, the most influential justice of the latter half of the twentieth century.[11] His especially forceful defense of textualism inspired a generation of conservative jurists while winning at least the grudging acceptance of many liberals. It even influenced proponents of the dominant liberal philosophy of jurisprudence,[12] which posited a "living Constitution" that might adapt in response to social pressure (yes, analogies to Darwinian ideas have been drawn and analyzed.)[13] Although she later retracted the comment, Justice Elena Kagan acknowledged, at a symposium focused on the influence of Scalia, that "we're all textualists now."[14]

The rise of formalist philosophies of interpretation entailed a retreat from the kind of permissive consideration of outside expertise that was symbolized by, among other things, the Brandeis brief. Many contests over evolution in particular involved the presentation of scientific evidence (and attempted rebuttal by anti-evolutionists). The involvement of Curtis and others at *Scopes*, Gould and others at *McLean*, and Gell-Mann and others at *Edwards* all come to mind.

Under modern theories of jurisprudence, however, science has grown more estranged from judging. Moreno has pointed to "false epistemic relativism" in which critiques grounded within the scientific method have been replaced with critiques *of* science.[15] In a 2017 case involving a dispute over election maps, a line of questioning by the conservative Chief Justice John Roberts focused on the challengers' empirical analysis of the impact of gerrymandering on voting power: "You're taking these issues away from democracy and you're throwing them into the courts pursuant to . . . sociological gobbledygook." The president of the American Sociological Association protested in a subsequent letter that the methods employed by the challengers were "rigorous and empirical," to no avail.[16]

Likewise, in the noted 2014 case *Burwell v. Hobby Lobby Stores*,[17] in which a business argued that it should not be required to provide contraceptive coverage to employees due to religious beliefs about abortion, the Court refused to engage with the argument that the drugs in question did not cause abortion because they block implantation, thus preventing the occurrence of pregnancy. Instead, the court simply held that "the businesses have religious objections to abortion, and according to their religious beliefs the four contraceptive methods at issue are abortifacients."

The influence of scientific evidence has declined even more in environmental cases.[18] In a major regulatory case involving climate change in 2022, one of many friend-of-the-court briefs was a classic Brandeis brief providing the Court with scientific expertise from leaders in the field. Contributors included professors from Harvard, Princeton, MIT, and Stanford; winners of the National Medal of Science (the highest American award for scientific achievement), the *Grande Médaille* (its French equivalent), and the Max Planck Research Award; and members of the National Academy of Sciences, the French Academy of Sciences, and the Royal Society.

This extraordinary group set forth in meticulous detail the clear consensus of the scientific community that human activity has unequivocally caused unprecedented global warming, that such warming has already affected every American and will have a drastic effect on daily life if not checked, and that the challenged regulatory program could make a difference.[19] The majority of the Supreme Court ignored them, striking down the plan of the Environmental Protection Agency (EPA) in a decision that did not once reference the brief or its conclusions. A subsequent feature in the prestigious scientific journal *Nature* warned of a Supreme Court that is "sceptical of—if not outright hostile towards—science."

The next term, a similar result was reached with respect to the Clean Water Act. In a challenge arguing for a narrow interpretation of that Act, which covers wetlands "adjacent to" navigable waters, the Court adopted virtually the narrowest possible interpretation. It determined that wetlands are only adjacent to navigable waters if they are continuously connected by visible surface water. Hydrologists and other independent scientists submitted vast evidence explaining why this interpretation does not reflect the scientific reality of waterways (which are often connected via groundwater, even if they lack a visible surface connection during much of the year). Again, the presentation of scientific expertise did not merely fail to win the case. It was not even taken up for discussion by the majority opinion, which turned in part on textualist distinctions between the definitions of the word "waters" (in the plural) compared with the word "water" (in the singular). This disinterest in science in regulatory cases was capped by the Court's formal disavowal of an influential doctrine (known as "Chevron deference," after a 1984 decision) that had heretofore extended formal legal weight to certain judgments by experts at regulatory agencies.[20]

This new formalism, and its epistemological disengagement with science, did not achieve dominance without resistance. Many liberal jurists initially pushed back against the core assumptions underlying textualism and originalism. The plain text of the Constitution is rarely a model of clarity, and the portions that are not subject to reasonable disputes over their meaning (two senators from every state, for example) are not the ones giving rise to cases that reach the Supreme Court. Liberals also chafed against the awkward task of determining fixed historical meaning in disputes over issues that did not even exist at the time of ratification. Thus, when Louisiana's new-model anti-evolution law reached the Court, near the beginning of the rise of originalism, Justice Brennan noted that free public education was "virtually nonexistent at the time the Constitution was adopted," rendering history "not useful in determining the proper roles of church and state in public schools."[21]

At that point, liberal Justices like Brennan could still rally a majority of the Court to strike down anti-evolution laws. But the easy near-unanimity that Fortas had enjoyed in *Epperson* had vanished, and the conservative legal movement was on the march. In particular, Scalia (who had recently joined the Court when Louisiana's law reached it) vigorously dissented against Brennan's decision:

> [W]e cannot say that on the evidence before . . . that 'creation science' is a body of scientific knowledge rather than revealed belief. *Infinitely less* can we say (or should we say) that the scientific evidence for evolution is so conclusive that no one could be gullible enough to believe that there is any real scientific evidence to the contrary.[22]

In a striking response to Scalia's dissent, Gould replied:

> [T]his is exactly what I, and all scientists, do say. We are not blessed with absolute certainty about any fact of nature, but evolution is as well confirmed as anything we know—surely as well as the earth's shape and position (and we don't require equal time for flat earthers and those who believe that our planet resides at the center of the universe).[23]

If these positions—the subject matter, tone, and argumentation—seem to echo 1925, it is no coincidence. Some things had changed since *Scopes*: *Epperson* and *Aguillard* remained binding precedent, controlling statements of law by the nation's highest court. But the midcentury consensus around those, and around the role of science as an appropriate guide for judicial decision-making, was gone. In some ways, the situation had reverted to 1925.

In *Science on Trial: The Case for Evolution*, evolutionary biologist Douglas J. Futuyma provides a superb account of the evidence for evolution while exposing various fallacies in the standard arguments used by opponents.[24] At the time he wrote the book, in 1983, Futuyma saw new attacks on evolution (and the resurrection of *Scopes*-era challenges) as a challenge to science itself. Although his work received excellent reviews, Futuyma did not ever expect to convert fundamentalist creationists to adherents of evolution. He felt that they would simply *not be swayed by evidence*. In the 1995 updated edition of *Science on Trial*, Futuyma expressed disappointment that there was an even greater need for the book at that point in time than when it was originally published:

> As we approach the 21st century, in an age in which some understanding of science is a virtual necessity for everyone, it is incredible that the single most fundamental principle of biology and one of the most fundamental in modern thought should still be an object of controversy and disbelief.

Incredible as it may be, the battle Darrow and Bryan undertook in 1925 remains relevant one hundred years later.

THE YEARS OF VAULTING AMBITION

These years saw even more internal scientific controversies among evolutionary biologists, which did nothing to counter the growing reluctance of courts to engage with the discipline (although legal scholars would participate). Much of the most heated controversy was rooted in the new field of evolutionary psychology, an extension and offshoot of the sociobiology movement begun by Edward O. Wilson.

Many historical accounts of evolutionary psychology point to its assuming disciplinary status in the late 1980s and early 1990s.[25] A core idea was that the cognitive mechanisms underpinning human behavior are adaptations, essentially, decision-making rules, or "Darwinian algorithms." Evolutionary psychologists posit that these rules enable organisms to detect and process evolutionarily relevant information from their physical and social environments, and respond with contextually and situationally appropriate behavioral responses, or "outputs."

The Darwinian algorithms themselves were not imagined to be directly observable through neuroscientific evidence. Instead, their form and design structure could be revealed through the generation of hypotheses: first, about the kinds of problems and challenges human ancestors faced, and then about what kinds of cognitive adaptations would have generated behavior that, on average, would have resulted in increased fitness. Thus, the *cognitive mechanisms are themselves the adaptations* for solving ancestral (Pleistocene-era) problems—but not necessarily problems encountered in current environments.

For evolutionary psychologists, the human mind therefore comprises a complex array of modular, domain-specific, cognitive problem-solving mechanisms. These modules are therefore central to any culturally independent and universal characterization of human nature. And a lot of this, naturally, has to do with sex.

Evolutionary psychology's initial foray into human sexual behavior was led by anthropologist Donald Symons, whose earlier research had examined play and aggression in rhesus monkeys. He switched gears in 1979, adopting a sociobiological perspective, and published *The Evolution of Human Sexuality*.[26] Through a slightly more formal evolutionary lens than had been applied in previous efforts (Desmond Morris in *The Naked Ape*,[27] for example) Symons explored the physical and psychological dimensions of sexuality, including reproductive anatomy, mate selection, sexual attraction, and emotional responses like jealousy. A central theme of the book was sexual selection, as initially conceptualized by Darwin, and its impact on human traits and behaviors.

Symons argued that, consistent with the animal literature on the subject, evolutionary pressures have resulted in distinct sexual strategies for human males and females. Men, he suggested, have evolved to pursue multiple mating opportunities to maximize reproductive output, given their lower parental investment. Women, he suggested, are inclined to be more selective in mate choice, reflecting their higher investment in offspring. Symons further explored the evolution of human pair bonding, hypothesizing its emergence from the need for extensive offspring care, leading to complex social and sexual behaviors such as monogamy. He also discussed the interplay between cultural factors and evolutionary biology in shaping human sexual behavior.

The book sparked considerable debate, especially regarding its interpretations of gender differences in sexuality. Critics raised concerns about missing relevant literature, the familiar issue of biological determinism, Symons' sporadic oversimplification of complex social behaviors (for example, he characterized copulation as a female "service," described female orgasm as epiphenomenal, and examined rape from a male reproductive perspective), and that the supporting empirical data for many claims and contentions in the book were, to say the least, sparse.[28]

But uproar elicited by Symons' book was minor compared with that which would be spawned by later work in the area of evolution and human sexuality. In the late 1980s, David Buss would take the reins and generate considerable interest and criticism with an expansive research program centered around the claim—rooted in sexual selection theory—that, cross-culturally, men seeking long-term partners tend to prioritize physical attractiveness and signs of fertility, whereas women value social stability and economic security more than men do.[29]

With little doubt, however, the most incendiary claims of evolutionary psychology centered around rape in humans.[30] In a 2000 book, *A Natural History of Rape: Biological Bases of Sexual Coercion*, biologist Randy Thornhill and anthropologist Craig T. Palmer argued that evolutionary theory provided insights about the occurrence of rape, which they suggested might be a side effect of evolved traits like sexual desire, aggression, and assertiveness—or, even more provocatively, an evolutionary adaption itself. Thornhill and Palmer argued that they were not committing the "naturalistic fallacy" with their proposal.[31] They contended that better understanding of rape would lead to more effective ways to reduce its occurrence, and they offered several recommendations in this light. But some of these recommendations, such as one involving women's choice of clothing,[32] only added to criticism that the book was insensitive and offensive. Thornhill and Palmer specifically challenged the idea, popularized by Susan Brownmiller in her 1975 book *Against Our Will*, that rape is purely an act of male power expression and intimidation, lacking sexual motives.

The book stirred considerable debate and media interest, especially after a segment was published in *The Sciences*. Reviewers (not just in science journals, but also prominent law reviews) were largely critical,[33] with feminist groups overwhelmingly opposing the notion that rape is a reproductive strategy. The authors were also challenged for distorting Brownmiller's views, for making dubious analogies between humans and other animals, and, again, for their suggestions to reduce the occurrence of rape. Critics strongly argued against the reproductive-adaptation theory of rape, citing many instances that did not fit this theory. The book was also condemned for its insensitive tone and emphasis on biological determinism.

Many scholars took issue with many of the claims of evolutionary psychology. Early on, in response to and in anticipation of critics, evolutionary psychology began contrasting itself with the purportedly traditional and decidedly oppositional view of the social sciences that human cognition is a general-purpose mechanism mostly constructed via culture.[34]

But internecine squabbles among evolutionists are often seen as the most damaging because what would otherwise be natural alliances can collapse as a result of them. Just as had been the case with sociobiology, many evolutionary biologists—

Gould and Lewontin were prominent exceptions—preferred to stay clear of the most contentious issues in evolutionary psychology by simply ignoring them.[35] Not, however, Jerry Coyne. When it comes to the idea that rape in particular is an adaptation; Coyne unabashedly said:

> If evolutionary biology is a soft science, then evolutionary psychology is its flabby underbelly . . . Once its scientific weaknesses are recognized, *A Natural History of Rape* becomes one more sociobiological "just-so" story—the kind of tale that evolutionists swap over a few beers at the faculty club. Such stories do not qualify as science, and they do not deserve the assent, or even the respect, of the public.[36]

Coyne claims that most of the public debates about *A Natural History of Rape* centered around ideology, but that the critical issue ought to have been about the scientific evidence behind the claims. And that is where the bulk of his criticisms lie. Coyne argues that Thornhill and Palmer's evidence is insufficient, biased, or as likely to be supportive of other explanations.[37] And Coyne was hardly alone in negatively assessing the evidence presented.[38]

Although Coyne and others acknowledge the potential value of evolutionary perspectives on human behavior, they stress the importance of empirical rigor and caution against the overextension of evolutionary theories to explain complex human social behaviors without sufficient evidence. There are, of course, echoes here of earlier attempts to apply, without sufficient evidence but with negative effects, evolutionary ideas to human societies. These did not end with Spencer and Galton.

These critiques are part of a broad and vigorous debate within the scientific community about the best ways to understand the evolutionary roots of human behavior, particularly behaviors with significant social and ethical implications. Of course, Thornhill and Palmer, and their supporters,[39] claim they are just following the science. And thus, the debate continues.[40]

For those in other disciplines, however, these fierce controversies served as a warning sign for researchers interested in evolutionary analyses of behavior. For example, legal scholar Owen D. Jones suggested that legal regimes dealing with child abuse, including infanticide, should incorporate insights about the potential evolutionary origins of these behaviors.[41] But in 1997, Jones's caution with respect to evolutionary psychology was palpable. He acknowledged that popular "notions of biological influences on human behavior comprise a patchwork of truths and untruths." Nevertheless, Jones challenged the legal scholarly discipline, in that "more than a century after Darwin's death, the very discipline designed to regulate human behavior often reflects, though perhaps unintentionally, the presupposition that all truly significant law-related human behavior is socially constructed." Jones then asserted that the behaviors that law regulates, including "aggression, risk-taking, deception, and sexuality," should not be divorced from their "evolutionary origins in the deep ancestral past." The extent to which evolution can explain human preferences for law-governed societies, as well as human tendencies to engage in the behaviors that law regulates, remains an intriguing area where legal and biological scholarship can intersect, even while the more extreme claims made by some evolutionary psychologists warn against overreach.

CHAPTER SIXTEEN

The End of *Lemon* and Calls to Revisit the Synthesis

> The anti-evolution law of the state of Tennessee, no matter whether it is repealed or not, will not be the only attempt to use the public schools as the effective purveyors of salutary political or religious truth.
>
> —"The Baiting of Judge Raulston," *New Republic*, July 29, 1925.

JUST TO THE NORTH OF RHEA COUNTY, Tennessee, sits Cumberland County, and just to the west of Cumberland County sits White County. Relatively poor and rural, the county seat of Sparta is smaller than Dayton, though the nearby rural counties are demographically similar.

White County never had a *Scopes* trial, but its legal department drew attention in 2017, when the county's general sessions judge, Sam Benningfield, issued a standing order that inmates who agreed to undergo a vasectomy procedure would receive thirty days of credit toward completion of their sentence.[1] Sign-up sheets and pamphlets were distributed to jails throughout the county. As a rural Tennessee trial judge, Benningfield had more in common with Judge Raulston than with Justice Holmes, but the sterilization-for-early-release order eerily evoked *Buck v. Bell*, issued during the height of the nation's interest in eugenics.[2]

Although it had originated with Galton, eugenics was not primarily or exclusively the province of evolutionists—and the White County order shows that the idea, and ones that are substantially similar to it, still linger long after eugenics was disavowed by all reputable scientists. Benningfield, however, defended his intentions, pointing to the number of children he had seen born with severe health problems arising from *in utero* drug exposure—a side effect of the monstrous opioid epidemic that swept through rural Tennessee and much of the rest of the country in the twenty-first century.[3] That epidemic, one of the largest public health crises in US history,[4] was particularly devastating to rural communities—and is almost certainly a contributing factor to why urban-rural cultural conflict in the twenty-first century feels harsher than it did to the participants at *Scopes*.

More challenges involving evolution and schools continued to arise in the early twenty-first century, but the tenor and makeup of the courts were gradually shifting. In 2000, Scalia and fellow conservative Clarence Thomas sarcastically decried a routine application of the Supreme Court's established *Epperson* and *Lemon* jurisprudence as pushing "the much beloved secular legend of the Monkey Trial one step further."[5] In Georgia, in 2005, litigation over whether a school district could place a sticker on its biology textbooks disclaiming that evolution "is not a fact" reached the Eleventh Circuit before settling in favor of the policy's challengers.[6]

Culture wars around science and religion continued to escalate, even after Peloza's *Scopes*-style challenge and the bitter "Scopes II" litigation arising in Rhea County. Another case billed as "Scopes II" arose in 2005, when a Dover, Pennsylvania, school district passed a resolution that "Darwin's Theory is a theory . . . Gaps in the Theory exist for which there is no evidence." The resolution was struck down by a federal district court through the straightforward application of the *Lemon* test.

However, the litigation became so contentious that the plaintiffs who had challenged the school district's actions received numerous death threats. So too did the presiding judge and his family, who had to receive additional protection from the US Marshals Service after receiving hate mail in the weeks following the trial.[7]

Nor was escalating hostility entirely one-sided. Many contributors to the *Scopes* legacy demonstrated no negativity toward religion itself: Asa Gray, Edward Loranus Rice, and Maynard Metcalf were all scientists who argued that evolution was consistent with their Christian faith. For his part, Gould consistently maintained that religion and science simply occupy separate spheres, each with few implications for the other. To Gould, "the best science often proceeds by putting aside the overarching generality and focusing instead on a smaller question that can be reliably answered . . . You might almost define a good scientist as a person with the horse sense to discern the largest answerable question—and to shun useless issues that sound bigger."[8] Others throughout this history, however, have taken more confrontational approaches. Darrow was never one to avoid confrontation for fear of causing offense. Mencken's attitude was even harsher. Indeed, Draper's thesis that religion and science stand in inherent tension, formulated long before *Scopes*, has never entirely fallen out of use.

After Gould died in 2002, the mantle of the most prominent public face of evolutionary biology passed to Richard Dawkins, whose credentials and contributions to the field are unimpeachable[9]. In 2006, Dawkins published *The God Delusion*,[10] presenting a grand historical clash between science and religion (and thereby echoing Draper). Rather than characterizing science and religion as existing in separate and reconcilable spheres (as Gould had), Dawkins engaged directly, forcefully responding to the argument, voiced long ago by Bryan and since by many others, that without a divine creator, life would lack purpose or moral direction. Instead, Dawkins argued that ethical systems do not require a religious foundation, that morality itself is a product of evolutionary processes, and thus that humanity can develop and foster a moral sense independent of religious teachings. He suggested that Darwinian evolution enables a different kind of meaning to life, one based on an understanding and appreciation of the natural world and universe. To Dawkins, this does not diminish human life but rather enriches it by providing a universal comprehension of our origins. Some readers—there have been many of them, with *The God Delusion* having sold more than three million copies—are convinced. But others responded much as Bryan might have, yet another piece of evidence suggesting that *Scopes* never truly faded from American consciousness.

Five years after the dispute in Dover, a New Jersey administrative law judge rejected an effort to place an anti-evolution referendum on an election ballot. The proposed referendum contended "that the theory of evolution is racist and communistic" and sought to ask voters whether they would "end the Theory of Evolution as Science and History in the presentation of the origin and duty of mankind in your Public Schools." The judge noted the referendum's similarity to *Scopes* and the Butler Act, holding that it was impermissible under *Lemon* and *Epperson*.[11]

More evolution debates arose in Ohio schools. In the early 2000s, the state board of education's benchmarks for grades nine and ten required children to describe "how scientists continue to investigate and critically analyze aspects of evolutionary theory." As we have shown, there is nothing inherently problematic about this language on its own; scientists do, in fact, continue to investigate, critically analyze, and vigorously debate aspects of evolutionary theory. But given the long historical context, one has to wonder the extent to which this language was formulated and applied in good faith. Regardless, the language was removed in 2006.

In response to its removal, an Ohio school teacher began distributing anti-evolution literature in the classroom and teaching children that evolutionary theory was wrong because, in his view, carbon dating is unreliable. (Here was direct evidence supporting Gell-Mann's assertion that anti-evolutionists frequently attempt to discredit other scientific fields.) The teacher was disciplined for failing to follow the curriculum, and in 2013, the Ohio state supreme court rejected his contention that this discipline violated his constitutional rights. But in the passionate dissents of three Ohio supreme court justices, the majority was criticized for ignoring the teacher's academic freedom and for purportedly telling "themselves that they are participating in the evolved version of the *Scopes* trial."[12]

And thus there have been significant evolution cases litigated in every completed decade between the 1960s and the 2010s. The decade that began in 2020 is unlikely to prove any exception. Indeed, during the drafting and editing of this book, the governor of West Virginia signed into law a new bill "allowing discussion of certain scientific theories" in public schools.[13] Without directly referencing intelligent design, creationist science, or other discredited anti-evolution labels, the act provides a safe harbor for teachers to address "questions from students about scientific theories of how the universe and/or life came to exist." The new law would seem to allow teachers to introduce creationism, so long as in response to a student question, and will certainly spark litigation. The legacy of *Scopes* marches on.

THE FALL OF *LEMON*

The conservative transformation of the federal judiciary became decisive after the 2016 election of Donald Trump, who was able to nominate a staggering three Supreme Court Justices in a single four-year term. (For comparison, Franklin Roosevelt, first elected in 1932, selected Hugo Black as his first nominee only in 1937.) Trump's appointments resulted in a six-to-three conservative supermajority and signaled the complete ascendency of conservative judicial philosophies. Each of Trump's appointed justices—especially Neil Gorsuch and Amy Coney Barrett—strongly identify as textualists. And they share a relatively faint commitment to the principle of *stare decisis* (the commitment of a court to abide by past decisions) at the level of the Supreme Court.[14]

The new majority acted quickly and boldly. In the space of two years, the Court overturned the constitutional right to an abortion, restricted the legality of gun control measures that it deemed not "consistent with the Nation's historical tradition," forbade the use of affirmative action in college admissions, mandated a dramatically limited interpretation of the Clean Water Act, and made several other consequential decisions.[15] Few areas of previous legal consensus seemed safe from potential revision.

This new reality spelled the end for the three-part *Lemon* test, which had led to the straightforward rejection of intelligent design and creationism in cases arising from Louisiana, Pennsylvania, Georgia, and even Dayton, Tennessee. In 2019, Justice Neil Gorsuch denounced the *Lemon* test as a "misadventure," and Justice Thomas argued that the First Amendment's prohibition against governmental establishment of religion should not be applied *under any circumstances* to acts of a state legislature. The *Lemon* precedent limped on, still occasionally applied by lower courts, until 2021. Then, in a case involving a public school football coach with a long history of leading students in prayer, a majority of the Supreme Court more explicitly overruled *Lemon* because it was "abstract" and "ahistorical." The majority also suggested that it would abandon "offshoot[s]" of *Lemon*. However, it remains to be seen which "offshoots" the justices had in mind. In dissent, Justice Sotomayor decried the Court's disservice to "our Nation's longstanding commitment to the separation of church and state."

The law undergirding the teaching of evolution in US schools now sits on a far less stable foundation than it has since the midcentury. For now, Fortas's opinion in *Epperson*, which predated *Lemon*, remains the binding precedent that controls the lower courts. But the modern Supreme Court's commitment to its own longstanding precedents has proven extremely fragile, as demonstrated by the rapid fall of both *Roe*, *Lemon*, the doctrine of "Chevron deference," and other foundational midcentury commitments. Should evolution reach the Supreme Court yet again, *Epperson* may be especially vulnerable given the similarities between Fortas's reasoning and the now-abandoned reasoning in *Lemon*, which has now been overruled. Precedents about the place of religion in public schools, particularly when religion is juxtaposed with the teaching of evolution, are now vulnerable across the board.

Ironically, the largest threat to the *Epperson* precedent now may *not* be the danger many scientists have warned about since the late 1970s, i.e., the various attempts by the anti-evolution movement to misrepresent internal scientific controversies and repackage fundamentalist objections as legitimate, alternative scientific theories. Instead, using analytical methodologies that afford primacy to text, original meaning, history, and tradition, the modern Court may simply decide that the Constitution has nothing to say about the proper place of science in the public curriculum. As Justice Brennan recognized in *Edwards*, public schools were "virtually nonexistent at the time the Constitution was adopted." In the gun-control context, another field marked by heated cultural controversy, the lack of sufficient, founding-era historical analogies has recently proved fatal to a variety of regulations.[16] A Court fully committed to analyzing the Constitution through the lens of founding-era history may not identify any meaningful restrictions on a state's legislative power to inject religion into public schools.

In another momentous trend, the modern Court has issued decisions authorizing greater public facilitation of religious *private* schools. In particular, the Supreme Court has struck down state laws that barred the public funding of religious education. For example, as a result of a recent decision, a Maine program that funds secular private schools is now constitutionally required also to fund religious schools, notwithstanding any state interest in promoting secular education.[17] And in 2024, a novel attempt to establish a "religious virtual charter school"—an instrument of the Catholic Church that would be chartered and funded by the state—was struck down by the Oklahoma Supreme Court.[18] But in dissent, one judge suggested that the decision was "destined" to be reversed by the Supreme Court.[19] Perhaps he was right. Either way, further movement in this area is likely to follow. Thus, just as the legal bars on religious instruction in public schools have weakened, so too are more students being pulled toward private education, where no such bars existed in the first place.

REVISITING THE SYNTHESIS

Notwithstanding renewed legal uncertainty regarding the place of evolution in the public school curriculum, there is today basic agreement that the central tenets of the Modern Synthesis remain the foundation for evolutionary biology. Little surprise, of course, that there is also broad acknowledgment within the field that the Synthesis remains "unfinished," in that uncertainty remains as to particular issues (some of which might seem of dubious concern to the non-specialist).[20] The last dozen years, however, have seen more substantial differences of opinion as to whether the traditional central principles of the Synthesis are indeed sufficient to address the full range of important questions emerging from current research.[21] In addition, many issues now being debated echo arguments that played out earlier in this history.

Some evolutionary biologists have advocated for what they call an "extended evolutionary synthesis" (EES) that would expand the recognized mechanisms underpinning evolution. Among the leaders of the group calling for this "rethink" is Kevin Lala (formerly known as Kevin Laland)[22] at the University of St. Andrews. The group shares the concern that the orthodox evolutionary view sees organisms as "simply programmed to develop by genes" passed down from ancestors—a modern critique that would have intrigued Darrow and resonated with Bryan.

This proposed anti "gene-centric" revision calls for an expansion of evolutionary mechanisms to include missing pieces of the puzzle. The press paid attention to these calls to action. In 2014, *Nature* published a "point-counterpoint" commentary titled "Does Evolutionary Theory Need a Rethink?"[23] *The Guardian*, not scientifically credentialed but nonetheless widely read, followed up in 2022 with an article titled, "Do We Need a New Theory of Evolution?"[24]

What exactly is missing from the Synthesis, according to these revisionists?

The first thing to observe about the current debates is that, unlike many of the others over the past hundred and fifty years,[25] these disagreements have remained civil and cordial—so far. Lala and his colleagues[26] acknowledge that the Modern Synthesis has been the dominant framework in evolutionary biology, with core components that include the randomness of genetic mutations that occur independent of natural selection's direction, gradual phenotypic change due to minor mutations, genetic inheritance, natural selection[27] as the sole driver of adaptation, and macroevolution as an accumulation of microevolutionary changes.

They contend, however, that although the Synthesis indeed expanded from its foundational structure in the mid-1950s (incorporating newer ideas like neutral theory, symbiogenesis, and inclusive fitness), it continues to view the processes of development and heredity separately, and tends to attribute undue causal and explanatory significance to genes. The group sees some newer research findings, in areas like evolutionary developmental biology, developmental plasticity, inclusive inheritance, and niche construction, as challenges to those traditional assumptions, and they have called for a reassessment of core evolutionary processes.

Supporters of EES argue for four additions to that orthodox framework. The first, "evolutionary developmental biology," merges critical aspects of evolutionary biology and developmental biology. The focus here is on how the development of organisms evolved, largely through genes that control developmental processes, and how changes in developmental processes can produce phenotypic variation in the natural world. Modifications in developmental pathways, it is argued, can lead to morphological and phenotypic diversity. Sources of bias in phenotypic variation are seen as important evolutionary processes that not only constrain (which is the traditional Synthetic view), but also facilitate and direct evolution.

This concept of "developmental bias" suggests that the way an organism develops from a fertilized egg into an adult can make certain physical traits more likely to emerge than others. Bias in development means that not all possible genetic changes have an equal chance of manifesting as physical traits. Some traits are more likely to appear because the organism's developmental processes are more receptive to changes that lead to these traits. This can significantly influence the direction and pace of evolution.

The second proposed extension concerns "developmental plasticity," referring to a capacity for phenotypes to be altered in response to environmental changes. This interpretation focuses on "how plasticity contributes to the origin of functional variation under genetic or environmental change and how the mechanisms of plasticity limit or enhance evolvability and initiate evolutionary responses."[28] The concepts of phenotypic and genetic accommodation—how organisms respond and adapt to environmental changes, both in the short term (phenotypic accommodation) and over evolutionary time (genetic accommodation)—are prominent in this perspective. Natural selection acts on genetic variations that underlie the phenotypically plastic traits. This leads to the evolution of new genetically stabilized traits that are better suited to the new environmental conditions.

To convey the idea through a relatable example, one can imagine humans newly inhabiting a high-altitude area. They are initially subject to lower oxygen levels than they are used to at sea level. In response to this environmental change, individuals may experience an increase in red blood cell production, enhanced lung capacity, and other physiological changes to accommodate the lower oxygen availability. This response is phenotypically accommodated—not due to genetic changes but rather an adjustment of the body's existing systems to the new environmental conditions. By way of genetic accommodation, the population gradually undergoes natural selection and shows genetic changes that reflect the initially phenotypically accommodated traits, making them a more permanent, and genetically heritable feature.

The third addition, "inclusive inheritance," as the term suggests, expands the focus of inheritance beyond DNA transmission. This broader view of heredity proposes that non-genetic inheritance can influence evolution (if the name Lamarck comes to mind again, it is with good reason). In the modern sense, this would include epigenetic[29] inheritance, which involves changes in gene expression that do not alter the DNA sequence but nonetheless can still be inherited. Epigenetic markers, influenced by environmental factors, can be passed from one generation to the next, affecting how genes are expressed. This broader notion of inheritance would also include some transmission of behavioral traits and patterns through learning and imitation, including culture (again, an updated but familiar refrain from earlier history).

Finally, the phenomenon of "niche construction" expands the orthodox Synthesis by recognizing that organisms actively modify their environments, or even create new environments, through their own behaviors and physiological activities. The environment, then, is not seen as a static entity that selects for or against certain traits in organisms. Instead, niche construction emphasizes a more dynamic interaction where organisms are both shaped by their environments and shape those environments in return.

Thus, the four components of the EES framework feature two unifying themes: constructive development and reciprocal causation. The former focuses on organisms shaping their developmental trajectories, emphasizing the interdependence of gene expression and the environment. Reciprocal causation posits that developing organisms are not just products of evolution but also partial drivers. EES thus recognizes that adaptation arises not only through natural selection but also via internal and external "constructive" processes. Proponents claim that while the EES does not necessitate a "revolution" in evolutionary biology, it represents more than just an extension of existing science; it requires, at the very least (they warn), conceptual change.

In response, there has been resistance and a failure to be convinced. Replying to the Lala *Nature* commentary, Gregory Allan Wray and colleagues contend that the four core expansions of the EES have, in fact, long been integrated into evolutionary biology, some tracing back to Darwin's time and his remarkable study of earthworm adaptation to soil (published in his last book), an early example of niche construction.[30] Phenotypic plasticity, they say, has for a long time attracted significant attention, evident in numerous studies documenting environment-induced trait variations. The traditionalists note that the role of plasticity in evolutionary change is well established, and that before any reassessment is required, its potential to *lead* genetic variation during adaptation remains to be clarified and better substantiated. This became a common theme in traditionalists' responses: "Been there, done that," and "Yes, perhaps, but we need more evidence." The traditionalists assert that while the concepts promoted by the EES are valuable areas to explore via the basic conventional evolutionary processes of natural selection, drift, mutation, recombination, and gene flow, they are not essential theorems for evolution and thus their status does not rise to a level that would require a major revision of the Synthesis.[31]

So, as *The Guardian* asked in 2022, "Does Evolutionary Theory Need a Rethink?" Well, stay tuned.

Here, and in the previous examples we have explored, the complex and variegated nature of evolutionary theory, and its complicated history, can be challenging for the public to fully grasp and decipher. Debates within the evolutionary biology community can be misconstrued (and misrepresented via disinformation) as disagreements about the validity of evolution itself. This misinterpretation creates confusion and skepticism, undermining the overwhelmingly accepted scientific consensus on evolution. This skepticism is particularly problematic in educational contexts, where it can influence how evolution is presented in schools.

Opponents of evolution have often seized upon these scientific debates as evidence against evolution. Disagreements among researchers are portrayed as fundamental flaws in evolutionary theory and used as ammunition—in the media and in court—to argue for alternative explanations. Many nonspecialists, even outside the highly motivated anti-evolutionist movement, routinely demonstrate inaccurate understandings of the history and science behind the theory of evolution. Recall, for example, California school authorities stating that traditional evolutionary theory did not encompass "changes in animal life,"[32] or Judge Lindsay's unfounded and incorrect assertion that Darwin believed human races evolved from different species of primates.[33] And as for anti-evolutionists themselves, they have used these debates to lobby for changes in educational curricula, advocating for the teaching of creationism or intelligent design alongside evolution in science classes. This push for "teaching the controversy" is misleading because it presents the scientific debate as a dichotomy between evolution and creationism, which is not an accurate reflection of the far more nuanced scientific discourse in the modern field.

THE RISE OF COVID-19 AND THE LONG LIFE OF *SCOPES*

As the COVID-19 pandemic engulfed the globe, killing several million people and nearly paralyzing the world's leading economies, many turned to modern science for hope. Vaccine science ultimately provided a solution many had prayed for. However, the development of effective vaccines merely brought about a new phase of humanity's relationship with the virus—one defined by Darwinian competition and adaptation. Now, scientists race to update COVID-19 vaccines each year to respond to newly developing variants, which in turn arise as adaptations to the selective pressures imposed by mass immunity.

As the nation struggled to win back control from the virus, a public assault on the legitimacy of vaccine science became a new, central battle in America's seemingly endless culture wars. Skeptics questioned, dismissed, and rebelled against the overwhelming scientific consensus that vaccines are safe and effective.

In fact, while these battles intensified during the pandemic, vaccine skepticism had already been rising beforehand. Years before COVID-19, the prescient scholar Joelle Anne Moreno had linked the rise of vaccine problems to the Supreme Court's failure to firmly distinguish between science and pseudoscience in *Edwards v. Aguillard*. She noted that, in a 2011 case involving a wave of claims brought under the National Childhood Vaccine Injury Act, the Court did not engage with, or even mention, the fact that the case arose against a backdrop of widespread belief in the validity of empirically rejected contentions asserting causal relationships between vaccines and autism, which drove a corresponding decline in vaccination use. And while the Centers for Disease Control (CDC) had announced that measles was eradicated from the United States in 2000, that finding was retracted in 2014, corresponding with declining rates of measles vaccine uptake.[34]

As with many cultural and scientific conflicts since the paradigmatic *Scopes* trial, these battles consistently found their way into courtrooms, particularly during the pandemic. A typical case was brought in Pennsylvania when several healthcare employees filed a civil rights suit challenging an employer mandate that they be vaccinated. The plaintiffs cast their objections to vaccines in a religious light, claiming religious discrimination. A typical objection read: "The Bible says that man has free will and I am using my free will, granted to me by God, to reject the vaccine. I have faith in my own immune system and the ability for my own body to heal itself."[35]

In a decision written by Judge Matthew Brann, a federal court in Harrisburg threw out the case. Brann wrote of the complaint:

> While cloaked in the language of religious discrimination . . . [the plaintiff's] complaint makes plain that she's after a modern-day Scopes trial. She believes that COVID-19 vaccines and tests are a hoax. And she wants this Court to vindicate her views.

Shortly thereafter, Brann found himself overseeing another litigation in New Jersey, brought by an unvaccinated attorney claiming his rights were violated by a temporary measure limiting courthouse access to those who could show either proof of vaccination or a recent negative COVID-19 test. The attorney sought to have Brann recused, arguing in part that his previous invocation of the *Scopes* trial indicated judicial bias against an anti-vaccine position.[36]

From a legal perspective, these complaints were, of course, meritless. But the easy manner in which they slotted into a continuing dialogue about *Scopes*, one hundred years after Dayton's grand spectacle formally came to a close, illustrates the trial's continuing hold on the nation's mind.

Conclusion

We embarked on this project because we believe that the *Scopes* trial, its background, and its impact on subsequent history offer lessons for other significant, complex, and contentious issues. Although we hope readers have reached their own conclusions from the panoramic history we have provided, here we offer a few of our own.

IMPLICATIONS FOR SCIENTISTS

Murray Gell-Mann's extensive involvement in the *Edwards v. Aguillard* litigation is revealing—and at first glance, perhaps surprising. Gell-Mann was a celebrated physicist whose transformative work on quarks and other elementary particles was not obviously or intuitively connected to Darwinian natural selection. Nevertheless, Gell-Mann clearly saw—and persuasively argued for—the relevance of evolutionary theory to other realms of science and society. Others have concurred. Philosopher Daniel Dennett speculates on the potential of evolutionary theory to influence future scientific and philosophical thought, particularly in understanding complex systems.[1] He suggests that evolution, with its emphasis on gradual, cumulative change, can provide valuable insights into tackling complex problems in various fields.

At the same time, the roadblocks that have stymied efforts to educate and foster acceptance of evolution in the United

States, as demonstrated by the *Scopes* legacy, have paralleled those facing other fields, including climate science. The scientific consensus and empirical evidence that Earth's climate is warming, and that this trend has been induced by an increase in atmospheric greenhouse gas concentrations linked to human activity, are overwhelming.[2] An empirical review of peer-reviewed publications randomly selected from the environmental science literature found only four skeptical climate-related papers out of three thousand sampled.[3] Nevertheless, public acceptance of this conclusion has been slow in coming, and climate science, like evolutionary theory, continues to face significant skepticism and resistance.[4]

In few areas has that resistance been felt with more impact than the court system. Although some state and foreign courts have issued helpful decisions,[5] federal courts have largely failed to engage with the problem, a maddening jurisprudential shrug several scholars have labeled "judicial nihilism."[6] And the U.S. Supreme Court's growing skepticism toward administrative expertise and disengagement from scientific evidence suggest that these trends are unlikely to change. The federal courts have been no better friend to those seeking climate mitigation than they were to progressive activists of Bryan's day.[7]

Historians Naomi Oreskes and Erik M. Conway have offered important insights into why climate scientists have faced such hurdles when attempting to explain the consensus in their field to the public: a world of science denial, and the tactics used by a small group of conflicted scientists to sow doubt and confusion about several critical issues—including not just climate change but also the health risks from tobacco smoke and acid rain.[8]

Oreskes and Conway's brilliant expose, *Merchants of Doubt*, revealed the strategies and motivations behind these efforts to distort scientific facts and the impact such disinformation has had on public policy and society. They identify a small group of scientists, each highly respected at early points in their respective fields, who became involved in promoting skepticism about scientific evidence on behalf of interest groups. Initially working with the tobacco industry, they employed tactics gleaned from public relations firms and from Cold War propaganda to cast doubt on the causal link between smoking and cancer. They then applied similar tactics to challenge the scientific consensus on acid rain, arguing that there was insufficient evidence to support proposed environmental regulation. The result was a delay in policy action to address important problems. Some of the same scientists, supported by think tanks and corporate groups, have played a key role in denying climate change. The strategies employed also take advantage of the ideological opposition to government regulation that is shared by significant numbers of Americans.[9]

Oreskes and Conway offer recommendations about how to combat this disinformation war, and it is here we see the strongest connections to the legacy of *Scopes*. They observe that the scientific community often faces intimidation from organized, sophisticated efforts aimed at discrediting their work, and this can incentivize silence or the avoidance of engagement with fallacious claims. "What's the point?" many ask. Skeptics will maintain their stance regardless of factual rebuttals, and debates may inadvertently validate the supposed "controversy" in the eye of the public. But Oreskes and Conway caution that neglecting broad public communication is outdated and indefensible in today's world. Instead, they urge scientists to acquire the skills to engage effectively and clearly with general audiences and the media.

Honesty and objectivity, although commendable, they add, often lead to the media and public focusing on any remaining uncertainties and caveats—the footnotes of research—that then overshadow core and well-founded conclusions. Scientists should confidently present conclusions when they are supported by robust evidence as facts. They also deem the historical knowledge of fields, often neglected by active researchers, as vital for effective communication with the public, especially when there is a need to counter false claims.

The role of journalists is essential, too. Enhanced understanding of the science itself will result in more accurate reporting. Oreskes and Conway are confident, however, that the scientific consensus about climate change "will eventually win public opinion."[10] We can hope that will indeed be the case—and that it will happen prior to the point where some critical mass of meteorological catastrophes has tipped the balance.

But lessons from the history of pubic engagement with evolution, and the legacy of the *Scopes* trial, are unfortunately sobering. Despite twists and turns since the publication of *Origin*, evolutionary theory has long been supported by massive empirical evidence. Even as researchers actively advance the field, evolution stands as a cornerstone of modern biology, providing a unified explanation for the diversity of life on Earth.

However, despite this strong scientific foundation, evolution has been and continues to be met with skepticism and outright denial. A 2019 Gallup poll revealed that 40% of Americans (and 23% of Americans with a college degree) still hold strictly creationist views of human origins (meaning the creation of humans in their present form within the last ten thousand years), highlighting the persistent challenge of gaining widespread acceptance of evolutionary theory.[11] As we have relayed, this resistance is rooted in a complex interplay of religious beliefs, perceived threats to moral values, misunderstandings about the nature of scientific theories, and cultural identification with a movement that has played a notable role in American life for over a hundred years. But these results are also strikingly similar to the number of Americans who continue to reject evidence of climate change, suggesting a core group within the public whose resistance to scientific conclusions runs deeper.[12]

Significantly, this continuing rejection of evolution is *not* due to any lack of appropriately targeted and clear information available to the public—bookstore shelves groan under the weight of such offerings. Nor is it due to a lack of eager and articulate scientists willing to engage with the public: names like Huxley and Gray, Rice and Curtis, Mayr and Dobzhansky, Gould and Dawkins, each with their own style and perspective, but each in their own way eager and skilled at addressing a wider audience. Thus, the remedies that Oreskes and Conway suggest, although helpful and completely rational, might not be enough.

Moreover, the complexity and abstract nature of both concepts add to the educational challenge. Much of the challenge involves accurately communicating the nature of scientific debate. The public has often failed to understand that debates within evolutionary biology are about the details of the process, not its overall validity. Emphasis on the nature of scientific inquiry and the robustness of scientific consensus is crucial. Educators must clarify how scientific theories, unlike hypotheses, are well-substantiated explanations based on a body of evidence. They must also highlight the difference between healthy skepticism (the need to be convinced of claims and contentions) and denialism (a refusal to acknowledge evidence). More broadly, educators must strive to convey the dynamic and self-corrective nature of science, emphasizing that debate and revision are integral parts of scientific progress.

Science is an intensely competitive human endeavor, and that is not going to change. For many senior biologists today, the first exposure to this fact came, not from textbooks, but from James D. Watson's *The Double Helix* (1968).[13] Watson provided a personal account of the scientific race to determine the structure of DNA. The book was controversial because of its vivid description of the intense competition that pervaded the scientific community hunting for the secret to heredity in the 1950s. It was surprising to many that science worked that way, particularly in a field with such profound implications as genetics and molecular biology.[14] The competition was not just a quest for knowledge, but also for prestige and recognition (very Darwinian indeed). To the extent that the public gains a greater understanding of the internal competitive dynamics within scientific fields, they may gain an improved ability to contextualize internal debates, such as the many we have examined (such as punctuated equilibrium, sociobiology, and EES).

Students benefit greatly when science is conveyed effectively. Our knowledge about evolution bestows on us far more than the self-satisfaction of having solved the puzzle of our existence. We, alone among all creatures that have ever existed on planet Earth, have figured out how we got to be here. Beyond this, evolution informs us about approaches to public health, agriculture, conservation, and numerous other fields: it continues to shape our world and we have the capacity to respond to its challenges if we apply our knowledge appropriately.[15]

IMPLICATIONS FOR LEGAL INSTITUTIONS

According to Gallup polls, the Supreme Court's approval rating in 2023—just 40%—was the lowest since the firm began collecting data in 2000. In that year, when the Court decided the intensely controversial *Bush v. Gore* decision, approval of the institution stood at 62%.[16] Legal scholars, whose profession is necessarily interested in the broad legitimacy of the legal system, have pointed to a variety of explanations for how and why the Court has fallen so far, including an increasing public recognition of the inherently political nature of its decision-making,[17] and the Court's failure to adopt a mandatory ethics code in the face of various allegations of impropriety by certain justices,[18] and a pattern of inflammatory and high profile confirmation hearings.[19] Whatever the cause, as the Supreme Court goes, so go the rest of the courts—confidence in the judicial system as a whole has fallen as well.[20]

It is unlikely that the Court's failure to engage with science in recent rulings is a direct and significant cause; as outlined above, large segments of the public themselves have often been unable or unwilling to grapple with areas of scientific consensus. Nevertheless, in the era of *Epperson*, the midcentury Court had earned the goodwill of the scientific community, an informed and influential group willing to engage with and affirm the legitimacy of judicial institutions. It is no coincidence that renowned scientists, from Curtis and Rice to Gould and Gell-Mann, were willing to devote so much time and effort to judicial education at various points in our story.

Today, the scientific community has become increasingly alienated from the judiciary in the wake of decisions that have ignored scientific evidence or evinced hostility toward scientific expertise. The prestigious peer-reviewed journal *Nature* ran an article in 2022 titled "The Supreme Court's War on Science"[21] and another in 2024 titled "Science versus Government: Can the Power Struggle Ever End?"[22] An increasing number of science writers have depicted the judiciary as a system where the interests of powerful political entities win out over evidence-based claims.[23]

This is not a state of affairs that the judiciary should welcome. The scientific community, made up of creative and sometimes brilliant thinkers who have dedicated their careers to expanding the boundaries of human knowledge, should be a natural ally to the legal system, which, in its own way, is dedicated to resolving disputes by evaluating and assessing the accuracy of claims. And although large groups of the lay public may resist scientific consensus, huge numbers of people admire the research community and take their lead from its conclusions. The growing alienation between courts and scientists does neither side any good.

IMPLICATIONS FOR US DEMOCRACY

And where, in this picture, do law and democracy fit? One lesson from the legacy of *Scopes* is that issues of scientific consensus and public education may impact many legal decisions in critical ways, but the reverse seems less true. Neither the ambiguous result at *Scopes* itself nor the string of decisive legal victories for evolution that followed the *Epperson* and *Lemon* decisions fundamentally changed the American public's relationship with evolutionary theory.

Here, Bryan's perspective was insightful. Elected officials representing constituent voters will always have a significant say in determining the curricula of public schools—as they must, in a democratic system. To be sure, there are constitutional limitations, such as those found in the Establishment Clause, and the vigorous enforcement of those limitations by dedicated attorneys like Clarence Darrow, as well as courageous litigants like John Scopes and Susan Epperson, is absolutely critical. But judges can only do so much, and they often prefer to do less.

The defense team at *Scopes* had set their sights on an appellate court; there was never real hope that they would win over the jury at Dayton. But for significant progress to be made, it is ultimately the general public that must, somehow, receive the message. Until then, the hundred years' trial will continue.

Epilogue

A New Political Landscape, November 2024

THE REELECTION OF DONALD TRUMP to serve a second, nonconsecutive term as president is a seismic event in American history, one likely to shape political and cultural discourse for years to come. Scholars and commentators have already begun exploring the connections between this event and the surge in anti-science and anti-establishment sentiments that gained momentum during the pandemic. For now, as of this writing, it is far too early to precisely identify the implications of this historical juncture for the themes discussed in this book, including the enduring rift between scientific consensus and political agendas. Nevertheless, one trend from Trump's first term seems certain to continue—and be magnified—in his second.

As we have discussed, one of the most significant developments of Trump's first term was the rightward shift of the federal judiciary. This shift led to, among other things, the abandonment of the *Lemon* test and an increasing receptiveness to religious participation in public schools. Even before Trump's re-election, states like West Virginia responded by exploring new approaches to classroom instruction on evolution. Others, like Louisiana, began introducing religion into classrooms in other ways, such as mandating displays of the Ten Commandments.[1] (As of this writing, that Louisiana law has been temporarily blocked by a Barack Obama–appointed district court judge. That decision is likely to be appealed to the highly conservative Fifth Circuit, and perhaps the Supreme Court.)

For all the ways in which a president's power is vast, their influence on the courts is largely a product of chance. Trump, in his first term, had the fortuitous opportunity to gain and fill three Supreme Court vacancies. Biden, in succeeding him, filled only one. Should Trump fill two more in his second term, a full majority of the Court will be made up of his appointees.

The lower courts matter as well, and judges appointed by Trump have pushed the envelope with respect to highly charged cultural issues like abortion. For example, a Trump-appointed district court judge pointed to a moribund 1873 law criminalizing the mailing of any "thing designed . . . for producing abortion," in a case challenging the FDA's authorization of abortion medication.[2] On appeal, another Trump-appointed judge wrote, "Doctors delight in working with their unborn patients—and experience an aesthetic injury when they are aborted."[3] That case remains pending as of this writing—and it remains unclear the extent to which that 1873 law, known as the Comstock Act, will be enforced going forward now that abortion is no longer constitutionally protected.

What is clear, however, is that as Trump is afforded the opportunity to appoint new judges in his second term, ongoing shifts in the role, temperament, and doctrines of the federal judiciary are likely to accelerate. The odds of *Epperson* being revisited and existing laws on evolution in schools being overturned have only increased. In an era when federal judges are openly contemplating the enforcement of an 1873 law criminalizing the mailing of products related to abortion, the laws of the 1920s that criminalized the teaching of evolution remain as relevant as ever.

ACKNOWLEDGMENTS

THE AUTHORS WOULD LIKE TO EXPRESS their deepest gratitude to those who have supported and guided *The Hundred Years' Trial* from its inception to its publication.

First, we extend our sincere thanks to Peter Tallack of The Curious Minds Agency, GmbH, whose enthusiasm and dedication made this project possible. We first met Peter at a writers' workshop sponsored by Emory University, where he encouraged us to pursue this work and provided essential support throughout the process.

Our sincere thanks to Laura Davulis, acquisitions editor at Johns Hopkins University Press, and Ezra Rodriguez, assistant acquisitions editor at JHUP, for their belief in this project and their careful guidance through the publishing journey. We are also grateful to the anonymous reviewers working with JHUP, whose thoughtful critiques and valuable suggestions greatly enriched the manuscript.

We owe special appreciation to Jacob Ellis, president of the Rhea County Historical & Genealogical Society, and Tom Davis, officer of the Society, for their assistance in uncovering pivotal historical details and granting us access to invaluable archival photographs. We are grateful for the trust the Society placed in our project, especially given past experiences with authors whose works, in the words of a colleague, amounted to "another Inherit the Wind hit job." We are deeply grateful that the Society, prioritizing historical accuracy over any single perspective, trusted us to approach the Scopes Trial with the nuance and rigor it deserves.

A special thanks goes to Deborah Bailey, our meticulous copy editor, and to Marguerite Roby, photograph archivist at the Smithsonian Institution Archives, for her assistance in locating and preserving the visual history that supports this work. We are also grateful to the Lawrence Family Trust for granting permission to quote from Inherit the Wind.

Finally, and more personally, we thank the family members who make our work possible. Alex extends a heartfelt thanks to Sam Xia, whose constant support enabled this project and so much else. And we both thank Dr. Sarah Gouzoules, who read the entire manuscript and provided extensive comments and suggestions. As wife and mother to the authors, Sarah's insights and understanding of our perspectives, as well as her keen editorial eye, were invaluable.

NOTES

Preface

1. Edward J. Larson, *Summer for the Gods: The Scopes Trial and America's Continuing Debate Over Science and Religion* (New York: Basic Books, 1997).
2. Randy Moore and Susan E. Brooks, *The Scopes Trial: An Encyclopedic History* (Jefferson, NC: McFarland, 2022); Randy Moore, *The Scopes "Monkey Trial": America's Most Famous Trial and Its Ongoing Legacy* (Santa Barbara, CA: Praeger, 2022); Randy Moore, *John Thomas Scopes: A Biography* (New York: Bloomsbury Academic, 2023).
3. Gregg Jarrett and Don Yaeger, *The Trial of the Century* (New York: Threshold Editions, 2023).
4. Brenda Wineapple, *Keeping the Faith: God, Democracy, and the Trial That Riveted a Nation* (New York: Random House, 2024).
5. Ronald L. Numbers, *The Creationists, Expanded Edition* (Cambridge, MA: Harvard University Press, 2006); Eugenie C. Scott, *Evolution vs. Creationism: An Introduction* (New York: Bloomsbury, 2008); Peter J. Bowler, *Monkey Trials and Gorilla Sermons: Evolution and Christianity from Darwin to Intelligent Design* (Cambridge, MA: Harvard University Press, 2009); Nicholas Spencer, *Magisteria: The Entangled Histories of Science & Religion* (New York: Simon and Schuster, 2023).

Introduction

1. For discussion of the legacy of *Marbury* (establishing judicial review) and *Plessy* (allowing segregation), see, e.g., Alexander M. Bickel, *The Least Dangerous Branch: The Supreme Court at the Bar of Politics* (New York: Bobbs-Merrill, 1962), 76.

2. Robert Shaffer, "Book Review: Lauri Lebo, *The Devil in Dover*," *Pennsylvania History: A Journal of Mid-Atlantic Studies* 76, no. 4 (2009): 519.
3. E.g., *Torres v. Bd. of Ed. of the City of Camden*, No. 101–5, 2010 WL 4105224 (EFPS Oct. 15, 2010); *Kitzmiller v. Dover Area Sch. Dist.*, 400 F. Supp. 2d 707 (M.D. Pa. 2005); *Doe v. Porter*, 188 F. Supp. 2d 904 (E.D. Tenn. 2002); *In re* Bundy, 852 F.3d 945 (9th Cir. 2017) (Gould, C.J., dissenting); *Marfork Coal Co. v. Callaghan*, 601 S.E.2d 55 (W.V. 2004) (Maynard, C.J., dissenting); 148 Cong. Rec. 12,604 (2002) (statement of Rep. Mike Pence).
4. E.g., *Juliana v. United States*, No. 6:15-CV-1517, 2017 WL 9249531 (D. Or. May 1, 2017); Douglas A. Kysar, "What Climate Change Can Do About Tort Law," *Environmental Law* 41 (2011): 1, 30.
5. E.g., *Berutti v. Wolfson*, No. 2:22-CV-04661, 2023 WL 1071624 (D.N.J. Jan. 27, 2023); *Mosher v. Marshall University*, No. 2021–1040, 2021 WL 11133132 (W. Va. Educ. St. Empl. Griev. Bd. Dec. 9, 2021). See also Max Boot, "Foes of Science Faced Ridicule at the Scopes Trial: We're Paying the Price 95 Years Later," *Washington Post*, July 8, 2020.
6. *Mozert v. Hawkins Cnty. Pub. Sch.*, 647 F. Supp. 1194 (E.D. Tenn. 1986); Francis J. Beckwith, "Science and Religion Twenty Years after *Mclean v. Arkansas*: Evolution, Public Education, and the New Challenge of Intelligent Design," *Harvard Journal of Law & Public Policy* 26 (2003): 455, 459.
7. The development of Scopes mythology is discussed in L. Maren Wood, "The Monkey Trial Myth: Popular Culture Representations of the Scopes Trial," *Canadian Review of American Studies* 32, no. 2 (2002): 147.
8. Jeffrey P. Moran, "Reading Race into the Scopes Trial: African American Elites, Science, and Fundamentalism," *Journal of American History* 90 (December 2003): 3; Andrew Nolan, "Making Modern Men: The Scopes Trial, Masculinity and Progress in the 1920s United States," *Gender and History* 19, no. 1 (2007); Tom Arnold-Forster, "Rethinking the Scopes Trial: Cultural Conflict, Media Spectacle, and Circus Politics," *Journal of American Studies* 56 (2022): 142; Adam R. Shapiro, *Trying Biology: The Scopes Trial, Textbooks, and the Antievolution Movement in American Schools* (Chicago: University of Chicago Press, 2013); Edward J. Larson, *Summer for the Gods: The Scopes Trial and America's Continuing Debate over Science and Religion*

(New York: Basic Books), 44–46; Edward J. Larson, "The Scopes Trial and the Evolving Concept of Freedom," *Virginia Law Review* 85 (1999): 503; Randy Moore, "Creationism in the United States: I. Banning Evolution from the Classroom," *American Biology Teacher* 60 (1998): 486, 491; Kevin P. Lee, "Inherit the Myth: How William Jennings Bryan's Struggle with Social Darwinism and Legal Formalism Demythologize the Scopes Monkey Trial," *Capitol University Law Review* 33 (2004): 347; Jonathan K. Van Patten, "The Trial of John Scopes," *South Dakota Law Review* 66 (2021): 273.

9. The trial transcript was rushed to publication: *The World's Most Famous Court Trial, Tennessee Evolution Case; A Complete Stenographic Report of the Famous Court Test of the Tennessee Anti-Evolution Act* (Cincinnati: National Book Co., 1925), hereinafter, "Transcript."
10. "Dramatic Scenes in Trial: Bryan Fixes Flood's Date and Defends Jonah and Joshua," *New York Times*, July 21, 1925. Fay Cooper-Cole, "A Witness at the Scopes Trial," *Scientific American* 120 (1959): 200.
11. *Scopes v. State*, 289 S.W. 363 (Tenn. 1927). Hereinafter, "Appeal."
12. "The Case Against America," *Times (London) Literary Supplement*, September 30, 1926, 638. For other examples of overseas coverage, see, e.g., "Americans Studying Evolution," *Observer* (London), July 19, 1925, 13; "The Crime of Teaching Evolution," *North China Herald & Supreme Court & Consular Gazette* (Shanghai), July 25, 1925, 62; "Pay Is Refused by Darrow; Got Money's Worth," *China Press* (Shanghai), August 7, 1925; "Evolution Trial: The Effort to Bar Expert Witnesses," *Manchester Guardian*, July 17, 1925, 12.
13. David A. Sellers, "The Circus Comes to Town: The Media and High-Profile Trials," *Law and Contemporary Problems* 71 (2008): 181.
14. William Jennings Bryan and Mary Baird Bryan, *The Memoirs of William Jennings Bryan* (Philadelphia: United Publishers of America, 1925), 61.
15. Jerome Lawrence and Robert E. Lee, *Inherit the Wind* (New York: Ballantine Books, 2007); *Inherit the Wind* (directed by Stanley Kramer, 1960).
16. For an illustrative and analogous contemporary controversy, see *Virden v. Crawford County*, No. 23-cv-2071, 2023 WL 5944154 (W.D. Ark. 2023), evaluating a claim that policies prohibiting public libraries from carrying children's books with LGBTQ themes entailed unconstitutional governmental establishment of religion.
17. 262 U.S. 390 (1923).

18. *McLean v. Arkansas Bd. of Ed.*, 529 F. Supp. 1255 (E.D. Ark. 1982).
19. *Kitzmiller v. Dover Area Sch. Dist.*, 400 F. Supp. 2d 707 (M.D. Pa. 2005).
20. See, e.g., *Juliana*, 2017 WL 9249531.
21. 393 U.S. 97 (1968).

Chapter One: An Inordinate Fondness for Beetles

1. "Probably no other major branch of science today is so haunted, dominated, and driven by the thoughts of one man." A poetic and accurate observation in: Jonathan Weiner, *The Beak of the Finch: A Story of Evolution in Our Time* (New York: Vintage Books, 2014), 128.
2. Earlier writers described species undergoing a process of "transmutation" or "transformation." That process would later become known as "evolution," a term that had previously appeared mostly in embryology. But even Darwin's *The Origin of Species* initially used "descent with modification." Only later editions adopted the term "evolution."
3. See David Sedley, *Creationism and Its Critics in Antiquity* (Los Angeles: University of California Press, 2008).
4. Two classic books, published around the hundredth anniversary of the *Origin*, are Loren Eiseley, *Darwin's Century: Evolution and the Men Who Discovered It* (Garden City, NY: Doubleday, 1958) and John C. Greene, *The Death of Adam* (Ames: Iowa State University Press, 1959).
5. There is a vast literature focused on Darwin, including some thirty book-length biographies, dating from shortly after his death to the present. Janet Browne, noted Darwin scholar and historian, suggests these works share, at their core, the exploration of the man's path to such enduring prominence. Browne emphasizes, however, that the tone, themes, and conclusions reached vary considerably depending on the time the biographies were written and the academic perspectives, backgrounds, and goals, of the authors. Modern biographers thus emphasize social and cultural conditions that shaped his actions, dealing "less with the details of the science and more with the cultural features that create a scientist." Janet Browne, "Making Darwin: Biography and the Changing Representations of Charles Darwin," *Journal of Interdisciplinary History* 40, no. 3 (2010): 347, 373; Janet Browne, *Charles Darwin: Voyaging* (New York: Knopf, 1995); and Janet Browne,

Charles Darwin: The Power of Place (New York: Knopf, 2002). By necessity, we are limited here to a concise, but aspiringly informative, account of how Darwin came to his ideas about evolution and why they caused such an enduring furor.

6. A little ashamedly, using the word "confess."
7. Even the word biology (*biologie*) was coined by Lamarck in his famous book *Philosophie Zoologique*, published in 1809, the year Darwin was born.
8. He would, however, later object to the idea that his success was due in part to evolutionary concepts being "in the air" at the time. Charles Darwin and Nora Barlow, *The Autobiography of Charles Darwin, 1809–1882: With Original Omissions Restored* (London: Collins, 1958).
9. This according to his former graduate student, John Maynard Smith. Haldane's exact words are debated, but the phrase is forever in the lexicon of evolutionary biology.
10. By all accounts the Reverend John Stevens Henslow (1796–1861) was an exceptionally kind and generous teacher to all students, but he took a particular liking to Darwin. Henslow's deep religious convictions never allowed him to accept Darwin's conclusions about evolution, but nonetheless he remained personally loyal to his former student. As a boy, Henslow had always wanted to explore the tropics and no doubt relished the prospect of taking on the job of ship's naturalist on the *Beagle*. But his wife did not allow it, setting the stage for him to recommend that Darwin serve in this capacity. Darwin and Henslow corresponded regularly during the entire five years of the voyage, with Henslow providing support and advice throughout. Francis Darwin, ed., *The Life and Letters of Charles Darwin*, Project Gutenberg, last modified March 14, 2019, https://www.gutenberg.org/cache/epub/2087/pg2087-images.html; Janet Browne, *Charles Darwin: Voyaging* (Princeton, NJ: Princeton University Press, 1995).
11. *Principles of Geology: Being an Attempt to Explain the Former Changes of the Earth's Surface, by Reference to Causes Now in Operation*, published in three volumes from 1830 to 1833.
12. Adrian Desmond and James Moore, *Darwin's Sacred Cause: Race, Slavery, and the Quest for Human Origins* (London: Penguin UK, 2009).
13. Not just for Darwin, but for the islands themselves, which are now the destination of some 170,000 tourists per year. The islands

continue to provide major insights into the workings of evolution. See Wiener, *supra*, who provides a superb account of the decades-long research program on Galápagos finches by husband-and-wife team Peter and Rosemary Grant. Weiner, *The Beak of the Finch*.

14. Norman A. Johnson, *Darwinian Detectives: Revealing the Natural History of Genes and Genomes* (Oxford: Oxford University Press, 2007). An excellent and readable account of how studies of genes and genomes have revealed evolutionary relationships among organisms.
15. John Van Wyhe, "Mind the Gap: Did Darwin Avoid Publishing His Theory for Many Years?" *Notes and Records of the Royal Society* 61, no. 2 (2007): 177. Wyhe concludes that "Darwin's delay is a recent historiographical theme for which there is no clear evidence, and indeed is overwhelmingly contradicted by the historical evidence." In other words, the "delay" in the publication of the *Origin* has perhaps been exaggerated, and the time to publication was to be expected given Darwin's goal of providing the most compelling case possible, and his other ongoing research projects.
16. Robert Chambers, *Vestiges of the Natural History of Creation*, ed. J. A. Secord (Chicago: University of Chicago Press, 1994), originally published 1844.
17. J. T. Costa, "Wallace, Darwin, and Natural Selection," in *An Alfred Russel Wallace Companion*, ed. Charles H. Smith, James T. Costa, and Michael P. Taylor (Chicago: University of Chicago Press, 2019), 97. The key difference was Darwin's focus on individuals in a population—to use modern terms, Wallace's account is more aligned with ideas of group selection. Theoreticians would later explore the evolutionary implications of the *unit of selection*—individual versus group—and come to understand that the circumstances under which selection might promote or eliminate groups (rather than individuals) would be far rarer.
18. The Linnean Society, "160th Anniversary of the Presentation of 'On the Tendency of Species to Form Varieties,'" July 1, 2018, https://www.linnean.org/news/2018/07/01/1st-july-2018-160th-anniversary-of-the-presentation-of-on-the-tendency-of-species-to-form-varieties.
19. Charles Darwin to J. D. Hooker, November 23, 1859, in *The Darwin Correspondence Project*, https://www.darwinproject.ac.uk/letter/DCP-LETT-2548.xml.

20. Six editions of *Origin* were published during Darwin's lifetime. The first (and the second, which largely involved minor corrections) best reflect Darwin's uncontaminated insights, because later editions included alterations that largely fail with respect to current evolutionary thought and were added to address the major contemporary criticisms the original book received. The leading "*On*" was dropped from the title in the sixth and final edition.
21. Peter J. Bowler, *Darwin Deleted: Imagining a World Without Darwin* (Chicago: University of Chicago Press, 2013).
22. Other scholars have mixed views, in principle, about the utility of such "counterfactual" histories. See A. C. Love, R. J. Richards, and P. J. Bowler, "What-If History of Science," *Metascience* 24, no. 1 (2015): 5–24.
23. Bowler, *Darwin Deleted*, 23.
24. "Wallace Letters Online: Letter no. 374," in The Natural History Museum, https://www.nhm.ac.uk/research-curation/scientific-resources/collections/library-collections/wallace-letters-online/374/374/T/details.html#4.

Chapter Two: One Long Argument Interrupted

1. There is debate as to when the famous nickname was first used, with some contention that it was not in Huxley's lifetime. John F. L. S. van Wyhe, "Why There Was No 'Darwin's Bulldog': Thomas Henry Huxley's Famous Nickname," *The Linnean* 1 (2019): 35.
2. T. H. Huxley to Charles Darwin (Nov. 23, 1859), *The Darwin Project*, https://www.darwinproject.ac.uk/letter/DCP-LETT-2544.xml.
3. The account of the 1860 Oxford debate that appeared in the July 14 edition of *Athenaeum* was, for many years, viewed as the most complete available, but more detailed documentation has more recently been discovered. Richard England, "Censoring Huxley and Wilberforce: A new source for the meeting that the *Athenaeum* 'wisely softened down,'" *Notes and Records: The Royal Society Journal of the History of Science* 71, no. 4 (2017): 371–384.
4. He also authored an important monograph on *Archaeopteryx*, the famous fossil of a bird-like winged creature with reptilian characteristics, after its discovery in 1860. In later editions of *Origin*, Darwin joyfully added reference to *Archaeopteryx* as a key piece of evidence for transitional evolutionary change.

5. "The Victorian Dinner Inside a Dinosaur," *The History Press*, December 20, 2016.
6. Richard Milner, *The Encyclopedia of Evolution: Humanity's Search for Its Origins* (New York: Facts on File, 1990), 236; Robert Peck, *All in the Bones: A Biography of Benjamin Waterhouse Hawkins* (Philadelphia: Academy of Natural Sciences, 2008).
7. Nicolaas A. Rupke, *Richard Owen: Victorian Naturalist* (New Haven, CT: Yale University Press, 1994).
8. Richard Owen, "Review of *On the Origin of Species* and Other Works," *Edinburgh Review* 111 (1860): 487–532, accessed January 8, 2024, https://darwin-online.org.uk/content/frameset?itemID =A30&viewtype=text&pageseq=1.
9. See Gregory A. Wickliff. "Draper, Darwin, and the Oxford Evolution Debate of 1860," *Earth Sciences History* 124 (2015): 35.
10. James C. Ungureanu, "A Yankee at Oxford: John William Draper at the British Association for the Advancement of Science at Oxford, 30 June 1860," *Notes and Records of the Royal Society of London* 69, no. 1 (2015): 83–99.
11. Richard England, "Censoring Huxley and Wilberforce: A New Source for the Meeting that the Athenaeum 'Wisely Softened Down,'" *Notes and Records of the Royal Society Journal of the History of Science* 71, no. 4 (2017): 371–384.
12. Wickliff, "Draper, Darwin, and the Oxford Evolution Debate," 140.
13. Charles Darwin to Charles Shaw (October 3, 1865), *The Darwin Project* https://www.darwinproject.ac.uk/.
14. Medical historian Howard Markel offers a detailed look at the personal and professional trials Darwin faced during the period surrounding the publication of *On the Origin of Species*. Markel explores how Darwin's health and the fluctuating support from his peers played crucial roles in his ability to navigate the contentious scientific and public reception of his work, emphasizing the intensity of the debates that followed its release. Howard Markel, *Origin Story: The Trials of Charles Darwin* (New York: W. W. Norton, 2024).
15. Darwin, *Origin* (1859), 282.
16. Philip C. England, Peter Molnar, Frank M. Richter, "Kelvin, Perry and the Age of the Earth: Had scientists better appreciated one of Kelvin's contemporary critics, the theory of continental drift might have been accepted decades earlier," *American Scientist* 95, no. 4 (2007): 342–349.

17. Darwin, *Origin* (1859), 308.
18. *Sw. Portland Cement Co. v. Reitzer*, 135 S.W. 237, 242 (Tex. Civ. App. 1911) ("The chain of causation too short. The connecting link is missing. And, like Darwin's 'missing link' in the evolution of man, it can't be found.").
19. Susan W. Morris, "Fleeming Jenkin and 'The Origin of Species': A Reassessment," *British Journal for the History of Science* 27, no. 3 (September 1994): 313–343; Michael Bulmer, "Did Jenkin's Swamping Argument Invalidate Darwin's Theory of Natural Selection?" *British Journal for the History of Science* 37, no. 3 (September 2004): 281–297.
20. Fleeming Jenkin, "Review of *The Origin of Species*," *North British Review* 46 (June 1867): 277–318, accessed January 12, 2024, https://darwin-online.org.uk/content/frameset?itemID=A24&viewtype=text&pageseq=1.
21. Jenkin's disturbing language, and the vile ideological assumptions that underlie it, are legitimate parts of the story of modern scientific development, in that opposition to racist scientific ideas about genetics and eugenics informed the actions of later anti-evolutionists like William Jennings Bryan. And yet, Jenkin's example also reveals that such ideas were common throughout the scientific community as a whole and by no means unique to or derived from evolutionary theory. Here, the racist imagery used by some contemporary scientists was deployed *against* Darwin's ideas, thus complicating later assertions (discussed later) about supposed links between evolutionary theory and racism.
22. Garrett Hardin, *Nature and Man's Fate* (New York: Holt, Rinehart & Winston, 1959).
23. Charles R. Darwin, *The Descent of Man, and Selection in Relation to Sex* (London: John Murray, 1871), 152, accessed January 20, 2024, https://darwin-online.org.uk/content/frameset?itemID=F937.1&viewtype=text&pageseq=1.
24. The basic idea is that altered characteristics, due to environmental effects, would be inherited by offspring; "quasi" in the sense that environmentally induced *random* changes can be beneficial and transmitted to offspring.
25. Ernst Mayr, *The Growth of Biological Thought: Diversity, Evolution, and Inheritance* (Cambridge, MA: Harvard University Press, 1982), 689–693.

26. The importance of use and disuse to Darwin are very clear in the *Origin*. This comes up repeatedly, especially in chapter 5, "Laws of Variation." He contends that the constant milking of cows leads to an inherited increase in the size of the udder. "I think there can be little doubt that use in our domestic animals strengthens and enlarges certain parts, and disuse diminishes them; and that such modifications are inherited." *Origin*, 134.
27. Thomas Henry Huxley, in typical fashion, jokingly cut to the core: "Genesis is difficult to believe, but Pangenesis is a deuced deal more difficult." T. H. Huxley to Charles Darwin (March 21, 1868), *The Darwin Project*, https://www.darwinproject.ac.uk/letter/DCP-LETT-6036.xml
28. Ernst Mayr, "Weismann and Evolution," *Journal of the History of Biology* 18, no. 3 (Autumn 1985): 295–329.
29. Weismann was among the first to tackle some of the most challenging questions in biology and evolution including sex, aging, and death, and his theories in these areas are largely consistent with modern thought. His empirical work included an infamous direct test of the inheritance of acquired traits that involved the amputation of mouse tails—over twenty generations—which, he was not surprised, had no effect on the tail length of offspring. There had been persistent anecdotes and lore to the contrary, prompting the experiment. Playwright and wit George Bernard Shaw ridiculed the experiment: ". . . any fool could have told him beforehand. Weismann then gravely drew the inference that acquired habits cannot be transmitted. And yet Weismann was not a born imbecile. He was an exceptionally clever and studious man, not without roots of imagination and philosophy in him which Darwinism killed as weeds" and "Weismann, a very clever and suggestive biologist who was unhappily reduced to idiocy by Neo-Darwinism." George Bernard Shaw, *Back to Methuselah*, Project Gutenberg, release date August 2, 2004, eBook #13084, accessed January 20, 2024, https://www.gutenberg.net/ebooks/13084.
30. Noteworthy in this discussion is the fact that cell theory—that cells are the basic units of all living tissues—was not proposed and supported until 1838, with the invention and rapid improvements of the microscope, and recognition that gametes are cells ("germ cells") came later, with sperm identified as cells by 1841 and eggs, finally, by 1861. It was not until 1875 that an understanding of

fertilization as the fusion of the nuclei of egg and sperm finally came about. Seemingly fanciful speculations of the sort represented by Darwin's concept of pangenesis should be seen in this context.

31. The concept of mutation, and an understanding of mutations and their role in evolution, would develop through convoluted paths over succeeding decades.
32. See Mayr, *The Growth of Biological Thought*, 701.
33. Until it reemerged in the form of epigenetic inheritance, a topic we return to in later chapters.
34. Their acceptance of the inheritance of acquired traits was largely consistent with Lamarck's, and thus the designation "Neo-Lamarckian." A concise review of Lamarck's historical relevance and contributions can be found in Richard W. Burkhardt Jr., "Lamarck, Evolution, and the Inheritance of Acquired Characters," *Genetics* 194, no. 4 (2013): 793–805.

Chapter Three: Survival

1. Janet Browne, "Asa Gray and Charles Darwin: Corresponding Naturalists," *Harvard Papers in Botany* 15, no. 2 (2010): 209–220.
2. Asa Gray, *Natural Selection Not Inconsistent with Natural Theology: A Free Examination of Darwin's Treatise On the Origin of Species, and of Its American Reviewers* (London: Trübner & Company, 1861).
3. Asa Gray, *Darwiniana: Essays and Reviews Pertaining to Darwinism* (Cambridge, MA: Harvard University Press, 1963).
4. Ernst Mayr, "Agassiz, Darwin, and Evolution." *Harvard Library Bulletin* 8, no. 2 (1959): 165–194.
5. A. Hunter Dupree, "The First Darwinian Debate in America: Gray Versus Agassiz," *Daedalus* 88, no. 3 (1959): 560–569.
6. Dupree, "The First Darwinian Debate in America."
7. Dupree, "The First Darwinian Debate in America."
8. Edward Lurie, "Louis Agassiz and the Races of Man," *Isis* 45, no. 3 (1954): 227–242.
9. Mark DeWolfe Howe, *Justice Oliver Wendell Holmes, The Shaping Years: 1841–1870* (Cambridge, MA: Belknap Press, 1957), 46, 52.
10. In the words of Richard Posner, "You can disagree with everything that Holmes ever said or thought yet regard him as a very considerable figure in American law because of his intellectual creativity and his rhetorical power; he altered the conception of the judge

and the structure of legal thought." Richard A. Posner, "In Memoriam: William J. Brennan, Jr.," *Harvard Law Review* 111 (1997): 9, 12.

11. Oliver Wendell Holmes Jr. to Morris R. Cohen, February 5, 1919, in Felix S. Cohen, "The Holmes-Cohen Correspondence," *Journal of the History of Ideas* 9, no. 1 (1948): 14–15.
12. Ronald L. Numbers, *The Creationists, Expanded Edition* (Cambridge, MA: Harvard University Press, 2006); Peter J. Bowler, *Monkey Trials and Gorilla Sermons: Evolution and Christianity from Darwin to Intelligent Design* (Cambridge, MA: Harvard University Press 2009); Ruth C. Stern and J. Herbie DiFonzo, "Dogging Darwin: America's Revolt Against the Teaching of Evolution," *Northern Illinois University Law Review* 36 (2016): 33.
13. Daniel C. Dennett, *Darwin's Dangerous Idea. Evolution and the Meanings of Life* (Simon & Schuster, 1995), 21.
14. William Jennings Bryan and Mary Baird Bryan, *The Memoirs of William Jennings Bryan* (Philadelphia: United Publishers of America, 1925), 535.
15. John W. Draper, *History of the Conflict Between Religion and Science* (New York: D. Appleton and Company, 1874).
16. Draper's historical work is no longer in vogue and has been criticized as infused with anti-Catholic bias. See Ronald L. Numbers, "Science and Religion," *Historical Writing on American Science* 1 (1985): 59–80. Yet works examining conflict between religion and science remain popular, if controversial. Prominent evolutionary biologist Richard Dawkins, for example, wrote one of several modern bestsellers about the implications of Darwinism for religion and morality, *The God Delusion* (London: Bantam Press, 2006), which sold more than three million copies.
17. *Eclipse Towboat Co. v. Pontchartrain R. Co.*, 24 La. Ann. 1, 13 (1872).
18. Andrew Carnegie, *The Gospel of Wealth* (New York: The Century Co., 1901), 3–4. See also John Hamer, "Money and the Moral Order in Late Nineteenth and Early Twentieth-Century American Capitalism," *Anthropological Quarterly* 21, no. 3 (Autumn 1998): 138.
19. Herbert Spencer, *The Principles of Biology*, vol. 1 (Cambridge: Cambridge University Press, 2011), originally published in 1864.
20. Diane Paul, "The Selection of the 'Survival of the Fittest,'" *Journal of the History of Biology* 21 (1988): 411.

21. Gregory Claeys, "The 'Survival of the Fittest' and the Origins of Social Darwinism," *Journal of the History of Ideas* 61, no. 2 (April 2000): 223.
22. Yair Sagy, "The Legacy of Social Darwinism: From Railroads to the 'Reinvention' of Regulation," *Georgetown Journal of Law & Public Policy* 11 (2013): 481.
23. E. Donald Elliott, "The Evolutionary Tradition in Jurisprudence," *Columbia Law Review* 85 (January 1985): 43–44.
24. Holmes read Spencer's *Social Statics* while recovering from wounds suffered at Chancellorsville. Kevin P. Lee, "Inherit the Myth: How William Jennings Bryan's Struggle with Social Darwinism and Legal Formalism Demythologize the Scopes Monkey Trial," *Capitol University Law Review* 33 (2004): 352–353.
25. Oliver Wendell Holmes Jr., "The Gas Stokers' Strike," *American Law Review* 7 (1873): 582.
26. Oliver Wendell Holmes Jr., "Law in Science and Science in Law," *Harvard Law Review* 12 (1899): 449.
27. Elliott, "The Evolutionary Tradition in Jurisprudence," 38.
28. 198 U.S. 45 (1905).
29. For example, in 1893, Harvard Law professor James Thayer outlined a theory that judges should strike down democratic legislation only when a constitutional failing was "so clear that it is not open to rational question," taking a strong view that—to put it colloquially—judges should give democratically enacted laws the benefit of the doubt. James B. Thayer, "The Origin and Scope of the American Doctrine of Constitutional Law," *Harvard Law Review* 7 (1893): 144.
30. Oliver Wendell Holmes Jr. to Harold Laski, July 23, 1925, in Richard Posner, ed., *The Essential Holmes: Selections from the Letters, Speeches, Judicial Opinions, and Other Writings of Oliver Wendell Holmes, Jr.* (Chicago: University of Chicago Press, 1992), 141. See also Richard A. Posner, "The Rise and Fall of Judicial Self-Restraint," *California Law Review* 100 (2012): 519.
31. G. Edward White, "Revisiting Substantive Due Process and Holmes's Lochner Dissent," *Brooklyn Law Review* 63 (1997): 87; Frederick Schauer, "Formalism," *Yale Law Journal* 97 (1988): 509.
32. Quoted in Hervert Hovenkamp, "The Marginalist Revolution in Legal Thought," *Vanderbilt Law Review* 46 (1993): 305.

Chapter Four: The Toilers Everywhere

1. Arthur Garfield Hays, *Let Freedom Ring* (New York: Boni and Liveright, 1928), 27–28.
2. John P. Altgeld, *Our Penal Machinery and Its Victims* (Chicago: A. C. McClurg & Co., 1886).
3. Ironically, some of the book's rhetoric echoes that of Bryan more than Darrow, e.g., "society demands protection to life and property and a preservation of the peace. That is all that it has any right to ask. It has no authority to sit in judgment on the sins of its members. This is a function which the Almighty has thus far reserved to Himself." Altgeld, *Our Penal Machinery*, 149.
4. Andrew E. Kersten, *Clarence Darrow: American Iconoclast* (New York: Hill and Wang, 2011), 30–31.
5. Scott Reynolds Nelson, "The Ordeal of Eugene Debs: The Panic of 1893, the Pullman Strike, and the Origins of the Progressive Movement," in *Workers in Hard Times: A Long View of Economic Crises*, ed. Leon Fink et al. (Champaign: University of Illinois Press, 2014), 99–110.
6. *United States v. E.C. Knight Co.*, 156 U.S. 1 (1895); *Pollock v. Farmers Loan and Trust Co.*, 158 U.S. 601 (1895); *In re Debs*, 158 U.S. 564 (1895).
7. William G. Ross, *A Muted Fury: Populists, Progressives, and Labor Unions Confront the Courts, 1890–1937* (Princeton, NJ: Princeton University Press, 1994), 10–13; Jane S. Schachter, "Putting the Politics of Judicial Activism in Perspective," *Supreme Court Review* 6 (2017): 218–221.
8. Kersten, *Clarence Darrow*, 84–90; Harvey Wish, "John Peter Altgeld and the Background of the Campaign of 1896," *Mississippi Valley Historical Society* 24, no. 4 (1938): 503.
9. William Jennings Bryan and Mary Baird Bryan, *The Memoirs of William Jennings Bryan* (Philadelphia: United Publishers of America, 1925), 101.
10. Bryan and Bryan, *Memoirs*, 496–514; Richard Franklin Bensel, *Passion and Preferences: William Jennings Bryan and the 1896 Democratic Convention* (Cambridge: Cambridge University Press, 2008), 203–47; Julian Harris, "Words that Thrilled: Bryan's Wonderful Sentence and Its Effect," *Atlanta Constitution*, July 2, 1896.
11. Kersten, *Clarence Darrow*, 110.

12. Louis D. Brandeis, to Alice Goldmark Brandeis, July 13, 1896, in *The Family Letters of Louis D. Brandeis*, ed. Melvin I. Urofsky and David W. Levy (Norman: University of Oklahoma Press, 2022), 82.
13. Clarence Darrow, *The Story of My Life* (New York: Charles Scribner's Sons, 1932), 92.
14. William Jennings Bryan, *First Battle: Story of the Campaign of 1896* (Chicago: W. B. Conkey Company 1896), 408. Bryan would repeat this slogan in later years. William Jennings Bryan, *The People's Law* (New York: Funk & Wagnalls Co., 1914), 55.
15. Louis D. Brandeis to Alice Goldmark Brandeis, November 28, 1899, in *The Family Letters of Louis D. Brandeis*, ed. Melvin I. Urofsky and David W. Levy (Norman: University of Oklahoma Press, 2022), 84.
16. Letter from William J. Bryan to Clarence Darrow (Mar. 29, 1902), in *The Clarence Darrow Digital Collection* (University of Minnesota Law Library); see also Harvey Wish, "Altgeld and the Progressive Tradition," *American History Review* 46, no. 4 (1941): 813, 817.
17. Theodore Roosevelt, "How Old Is Man?" *National Geographic Magazine* 29, no. 2 (February 1916): 114.
18. *Dayton Coal & Iron Co. v. Barton*, 53 S.W. 970 (Tenn. 1899), *aff'd*, 183 U.S. 23 (1901).
19. *Dayton Coal & Iron Co. v. Barton*, 183 U.S. 23, 23 (1901).
20. *Weaver v. Palmer Bros.*, 270 U.S. 402 (1926); *Adkins v. Children's Hospital*, 261 U.S. 525 (1923); *Hammer v. Dagenhart*, 247 U.S. 251 (1918).
21. *Ives v. S. Buffalo Ry. Co.*, 94 N.E. 431, 436 (1911).
22. Clifford Thorne, "Will the Supreme Court Become the Supreme Legislator of the Land?" *American Law Review* 43 (1909): 263.
23. Bryan, *The People's Law*, 53–54.
24. Bryan, *The People's Law*, 53–54.
25. Richard H. Pildes, "Forms of Formalism," *University of Chicago Law Review* 66 (1999): 608–609; Lyrissa Barnett Lidsky, "Defensor Fidei: The Travails of a Post-Realist Formalist," *Florida Law Review* 47 (1995): 819.
26. Thomas C. Grey, "Langdell's Orthodoxy," *University of Pittsburgh Law Review* 45 (1983): 5–11.
27. For example, he argued that for those studying law, "the man of the future is the man of statistics and the master of economics. It is revolting to have no better reason for a rule of law than that so

it was laid down in the time of Henry IV. It is still more revolting if the grounds upon which it was laid down have vanished long since, and the rule simply persists from blind imitation of the past." Oliver Wendell Holmes Jr., "The Path of the Law," *Harvard Law Review* 10 (1897): 469.

28. Oliver Wendell Holmes Jr., *The Common Law* (Boston: Little, Brown & Co., 1881), 35.
29. Oliver Wendell Holmes Jr., "Law in Science and Science in Law," *Harvard Law Review* 12 (1899): 462.
30. 208 U.S. 412, 419 n. 1 (1908).
31. Although some decisions before *Muller* did take peripheral data into account, Brandeis's integration of data and law represented clear innovation and a break with the predominant formalist philosophy, which left scant room for scientific insights in legal decision making. Martha Minow, "Justice Engendered," *Harvard Law Review* 101 (1987): 88–89; Michael Rustad and Thomas Koenig, "The Supreme Court and Junk Social Science: Selective Distortion in Amicus Briefs," *North Carolina Law Review* 72 (1993): 104; Noga Morag-Levine, "Facts, Formalism, and the Brandeis Brief: The Origins of a Myth," *University of Illinois Law Review* 2013, 60.
32. Andrew Berry and Janet Browne, "Mendel and Darwin." *Proceedings of the National Academy of Sciences* 119, no. 30 (2022): e2122144119. This article also nicely summarizes the rediscovery and interpretation of Mendel's work at the turn of the century.
33. In 1865, Mendel presented the results of his work at a meeting of the Natural History Society of Brünn, but there was little interest in them. He published them the next year in *Proceedings of the Natural History Society of Brünn*. G. J. Mendel, "Experiments on Plant Hybridization [Versuche über Pflanzen-Hybriden]," *Proceedings of the Natural History Society* 119 (1865): 3–47; N. C. Stenseth, L. Andersson, and H. E. Hoekstra, "Gregor Johann Mendel and the Development of Modern Evolutionary Biology," *Proceedings of the National Academy of Sciences* 119, no. 30 (2022): e2201327119, https://doi.org/10.1073/pnas.2201327119.
34. The "phenotypic ratio."
35. Specifically, an organism with two identical alleles for a gene is said to be homozygous for that gene (and is called a homozygote). An organism that has two different alleles for a gene is said to be heterozygous for that gene (and is called a heterozygote).

36. It later became clear that although dominance of one trait over another is quite common, it is not a universal phenomenon in inheritance. For some traits, individuals in the first generation are intermediate between the parental types (a form of "non-Mendelian inheritance").
37. The law of independent assortment was later modified in the early twentieth century to account for "linkage," which is when genes in close proximity on a chromosome have a greater likelihood of being inherited together.
38. William B. Provine, *The Origins of Theoretical Population Genetics* (Chicago: University of Chicago Press, 1971), 83–91; Garland E. Allen, *Thomas Hunt Morgan: The Man and His Science* (Princeton, NJ: Princeton University Press, 1978), 135–42; Peter J. Bowler, *Evolution: The History of an Idea* (Berkeley: University of California Press, 2009), 264–73.
39. Michael Bulmer, *Francis Galton: Pioneer of Heredity and Biometry* (Baltimore: Johns Hopkins University Press, 2003).
40. Peter J. Bowler, *The Mendelian Revolution: The Emergence of Hereditarian Concepts in Modern Science* (Baltimore: Johns Hopkins University Press, 1989); Bulmer, *Francis Galton: Pioneer of Heredity and Biometry*.
41. It was embraced, for example, by the French comparative anatomist Étienne Geoffroy Saint-Hilaire in the 1830s.
42. Darwin, *Origin of Species*, 471.
43. Along with Carl Correns and Erich von Tschermak.
44. Peter J. Bowler, *The Eclipse of Darwinism: Anti-Darwinian Evolution Theories in the Decades around 1900* (Baltimore: Johns Hopkins University Press, 1983), 40.
45. Stephen Jay Gould, *The Structure of Evolutionary Theory* (Cambridge, MA: Belknap Press, 2002), 451–66.
46. It would not be appropriate or possible to attempt a summary here of the vast scholarship on eugenics (there are many such reviews available), but some basic details and background is relevant to later discussions, especially for readers relatively unfamiliar with the area. For further sources, see, for example, Daniel J. Kevles, *In the Name of Eugenics: Genetics and the Uses of Human Heredity* (Berkeley: University of California Press, 1985); Mannie Liscum and Michael L. Garcia, "You Can't Keep a Bad Idea Down: Dark History, Death, and Potential Rebirth of Eugenics," *Anatomical Record* 305, no. 4

(2022): 902–937. Additional reviews cited in Christine M. Shea, review of "In the Name of Eugenics: Genetics and the Uses of Human Heredity" by Daniel J. Kevles, *History of Education Quarterly* 26, no. 4 (1986): 621–626.

47. Galton, a child prodigy, was a prolific writer and made significant contributions to genetics, statistics, psychology, and even far-flung fields such as meteorology. In his investigations of how environment and heredity contribute to particular traits like intelligence, Galton coined—with perhaps some help from Shakespeare—and popularized the phrase "nature and nurture."
48. Francis Galton, *Hereditary Genius: An Inquiry Into Its Laws and Consequences* (New York: Macmillan, 1869).
49. Derived from the Greek "*eu*" (εὖ), meaning "good" or "well," and "*genes*" (γενής), meaning "born" or "of a certain kind."
50. Francis Galton, *Memories of My Life* (Oxford: Routledge, 2015), chapter 21.
51. Ernst Mayr, *The Growth of Biological Thought: Diversity, Evolution, and Inheritance* (Cambridge, MA: Harvard University Press, 1982), 62–63.
52. Oliver Wendell Holmes Sr., "Crime and Automatism," *Atlantic Monthly*, April 1875, 475.
53. University of Missouri Libraries, "Controlling Heredity," accessed January 20, 2024, https://library.missouri.edu/specialcollections/exhibits/show/controlling-heredity/mizzou.
54. George William Hunter, *A Civic Biology: Presented in Problems* (New York: American Book Company, 1914), accessed January 8, 2024, https://www.gutenberg.org/files/39969/39969-h/39969-h.htm#Page_261.
55. Henry Herbert Goddard, *The Kallikak Family: A Study in the Heredity of Feeble-Mindedness*. (New York: Macmillan, 1912). R. L. Dugdale's 1877 account of the Jukes family was updated in Arthur Howard, "The Jukes in 1915," *Carnegie Institution of Washington*, no. 240 (1916).
56. P. A. Lombardo, *Three Generations, No Imbeciles: Eugenics, the Supreme Court, and Buck v. Bell* (Baltimore: Johns Hopkins University Press, 2011).
57. *Buck v. Bell*, 274 U.S. 200 (1927).
58. Bryan and Bryan, *Memoirs*, 548.

59. Arthur Garfield Hays, *Trial By Prejudice* (New York: Covici-Friede, 1933), 355–356.
60. Clarence Darrow, *Crime: Its Cause and Treatment* (New York: Thomas Y. Crowell Co., 1922), 253–54.
61. As we will see, debate and controversy persist.
62. Michael Kazin, *A Godly Hero: The Life of William Jennings Bryan* (New York: Alfred A. Knopf, 2006), 263; Kevin P. Lee, "Inherit the Myth: How William Jennings Bryan's Struggle with Social Darwinism and Legal Formalism Demythologize the Scopes Monkey Trial," *Capitol University Law Review* 33 (2004).
63. William Jennings Bryan, *In His Image* (New York: Fleming H. Revell Co., 1922), 15.

Chapter Five: Divergence

1. Ruth C. Stern and J. Herbie DiFonzo, "Dogging Darwin: America's Revolt Against the Teaching of Evolution," *Northern Illinois University Law Review* 36 (2016).
2. Woodrow Wilson, *Constitutional Government in the United States* (New York: Columbia University Press, 1908): 54–55.
3. Woodrow Wilson to Winterton Curtis, August 29, 1922, in W. C. Curtis, *Scopes Trial Scrapbook* (University of Missouri Archives).
4. Edward S. Mihalkanin, *American Statesmen: Secretaries of State from John Jay to Colin Powell* (Westport, CT: Greenwood Press, 2004) 74–81.
5. William Jennings Bryan and Mary Baird Bryan, *The Memoirs of William Jennings Bryan* (Philadelphia: United Publishers of America, 1925), 353.
6. Bryan and Bryan, *Memoirs*, 352. Holmes did, eventually, read Gibbon's *Decline and Fall of the Roman Empire*. See Oliver Wendell Holmes Jr. to Harold Laski, March 11, 1922, in Richard Posner, ed., *The Essential Holmes: Selections from the Letters, Speeches, Judicial Opinions, and Other Writings of Oliver Wendell Holmes, Jr.* (Chicago: University of Chicago Press, 1992).
7. 18 U.S.C. §§ 792–799.
8. A federal court went so far as to order the seizure of all copies of a film about the American Revolution, titled "The Spirit of '76," because it portrayed British soldiers in a negative light and might create "animosity or want of confidence between us and our allies." *United States v. Motion Picture Film "The Spirit of '76,"* 252 F. 946, 948 (S.D. Cal. 1917).

9. Woodrow Wilson to Clarence Darrow, August 9, 1917, in *The Clarence Darrow Letters*, https://librarycollections.law.umn.edu/darrow/letters/letter_detail.php?id=438.
10. *Debs v. United States*, 249 U.S. 211, 215 (1919).
11. Mark DeWolfe Howe, *Justice Oliver Wendell Holmes, The Shaping Years: 1841–1870* (Cambridge, MA: Belknap Press, 1957), 69–173. Holmes expounded on the themes of war and patriotism in his memorial day address to the graduating class at Harvard, "The Soldier's Faith," in Posner, ed., *The Essential Holmes*, 87–94.
12. See *Sullivan v. Flannigan*, 8 F.3d 591, 595 n. 3 (7th Cir. 1993).
13. Eugene Debs to Clarence Darrow, February 1, 1922, in *The Clarence Darrow Letters*, https://librarycollections.law.umn.edu/darrow/letters/letter_detail.php?id=438.
14. Burl Noggle, *Into the Twenties: The United States from Armistice to Normalcy* (Champaign: University of Illinois Press, 1974), 113.
15. *Abrams v. United States*, 250 U.S. 616, 630 (1919).
16. Ari Ezra Waldman, "The Marketplace of Fake News," *University of Pennsylvania Journal of Constitutional Law* 20 (March 2018): 847 ("Good ideas, like the best products, will win out and bad ideas, like inferior, faulty, or poorly made products, will be tossed aside.").
17. Oliver Wendell Holmes Jr., "Law in Science and Science in Law," *Harvard Law Review* 12 (1899): 449.
18. Vincent Blasi, "Holmes and the Marketplace of Ideas," *Supreme Court Review* 2004, no. 1 (2004): 24.
19. Benjamin Gitlow, *I Confess: The Truth About American Communism* (New York: E. P. Dutton & Co., Inc., 1940), accessed January 8, 2024, https://archive.org/stream/BenjaminGitlow/BenjaminGitlow_djvu.txt.
20. *Gitlow v. New York*, 268 U.S. 652, 672 (1925).
21. 262 U.S. 390 (1923).
22. In Bryan's memoirs, posthumously completed by his wife, the authors connected the *Meyer* case to the anti-evolution movement: "the United States Supreme Court rendered a decision emphasizing the parents' interest in the child's religion and affirming the state's right to control the schools. Legislatures here and there began to take notice." Bryan and Bryan, *Memoirs*, 480. Their reading of the case was strained; it nevertheless shows that Bryan was paying attention.

23. St. Louis Post-Dispatch, *Scrapping Bryan*, July 7, 1923, 10.
24. See generally Edward J. Larson, *Summer for the Gods: The Scopes Trial and America's Continuing Debate over Science and Religion* (New York: Basic Books, 1997). As noted, analysis of theological debates related to evolution is largely beyond the scope of this work but has been covered by others, including Larson and Numbers.
25. Advertisement, "Hell and the High Schools, and Other Books by T. T. Martin for Sale Here," in W. C. Curtis, *Scopes Trial Scrapbook*.
26. Mayr notes that the term "Mendelism" has different connotations. To some it refers to the period during which Mendel's work concerning the particulate nature of inheritance was rediscovered, confirmed, and furthered. In the other sense, it refers to the period when mutationism was promoted as an alternative to Darwinian natural selection. Bateson was involved in both senses. Ernst Mayr, *The Growth of Biological Thought* (Cambridge, MA: Belknap Press, 1982), 732.
27. "Scientist Denies a Darwin Theory," *New York Times*, August 15, 1914, 8.
28. William Jennings Bryan, "God and Evolution: Charge that American Teachers of Darwinism 'Make the Bible a Scrap of Paper,'" *New York Times*, February 26, 1922, 84.
29. W. C. Curtis to William Bateson, November 25, 1922, in University of Missouri Archives, W. C. Curtis, *Scopes Trial Scrapbook*.
30. W. C. Curtis to Woodrow Wilson, August 25, 1922, in University of Missouri Archives, W. C. Curtis, *Scopes Trial Scrapbook*.
31. Woodrow Wilson to W. C. Curtis, August 29, 1922, in University of Missouri Archives, W. C. Curtis, *Scopes Trial Scrapbook*.
32. Bryan and Bryan, *Memoirs*, 533.
33. "W.G.N. Put 'On Carpet'; Gets a Bryan Lashing: Commoner Assails Its Editorials," *Chicago Daily Tribune*, June 20, 1923.
34. "Darrow Asks W. J. Bryan to Answer These: For Instance: 'Did Noah Build Ark?,'" *Chicago Daily Tribune*; July 4, 1923.
35. "The Amazing and Amusing Mr. Bryan," *Chicago Daily Tribune*, June 21, 1923, 8.
36. W. B. Norton, "Bryan Brushes Darrow Bible Queries Aside: Says His Quarrel 'Isn't with Atheists,'" *Chicago Daily Tribune*, July 5, 1923.
37. Hal Higdon, *Leopold and Loeb: The Crime of the Century* (Urbana: University of Illinois Press, 1975); Simon Baatz, *For the Thrill of It:*

Leopold, Loeb, and the Murder That Shocked Chicago (New York: HarperCollins, 2008).

38. Compare with Holmes in "The Path of the Law": "If the typical criminal is a degenerate, bound to swindle or to murder by as deep seated an organic necessity as that which makes the rattlesnake bite, it is idle to talk of deterring him by the classical method of imprisonment." Oliver Wendell Holmes Jr., "The Path of the Law," *Harvard Law Review* 10 (1897).
39. Darrow, *The Story of My Life*; Scott W. Howe, "Reassessing the Individualization Mandate in Capital Sentencing: Darrow's Defense of Leopold and Loeb," *Iowa Law Review* 79 (1994): 989, 994–1036.
40. *Florida v. Nixon*, 543 U.S. 175 (2004); *Wade v. Calderon*, 29 F.3d 1312 (9th Cir. 1994) (Reinhardt, J., dissenting in part); *Resnover v. Pearson*, 754 F. Supp. 1374 (N.D. Ind. 1991).
41. Robert M. La Follette to Clarence Darrow, April 1, 1925, in *The Clarence Darrow Letters*, https://librarycollections.law.umn.edu/darrow/letters/letter_detail.php?id=358
42. *Tyree v. Kentucky*, 279 S.W. 990 (1926).
43. Bryan and Bryan, *Memoirs*, 546–47.
44. "The Evolutionary Debate at Dartmouth," *Dartmouth Undergraduate Journal of Science*, May 22, 2009, accessed May 18, 2024, https://sites.dartmouth.edu/dujs/2009/05/22/the-evolutionary-debate-at-dartmouth/.
45. Stuart Rice and Malcolm Willey, "Dartmouth Charts Mr. Bryan's Arguments," *New York Times*, January 13, 1924.
46. Edward L. Rice, "Darwin and Bryan—A Study in Method," *Science* 61, no. 1575 (1925): 243–250.
47. Bryan and Bryan, *Memoirs*, 533.

Chapter Six: A Magnificent Opportunity to Test an Obnoxious Law

1. "Fights Evolution to Uphold Bible: Author of Tennessee Law Tells Motives That Led Him to Frame It," *New York Times*, July 5, 1925.
2. Ruth C. Stern and J. Herbie DiFonzo, "Dogging Darwin: America's Revolt Against the Teaching of Evolution," *Northern Illinois University Law Review* 36 (2016): 35.
3. Adam R. Shapiro, *Trying Biology: The Scopes Trial, Textbooks, and the Antievolution Movement in American Schools* (Chicago: University of Chicago Press, 2013), 81–82.

4. "Fights Evolution to Uphold Bible: Author of Tennessee Law Tells Motives That Led Him to Frame It," *New York Times*, July 5, 1925.
5. A. G. Keller, "Law in Evolution," *Yale Law Journal* 28 (June 1919): 769.
6. Keller, "Law in Evolution," 775.
7. "Urges Retaliation on Evolution Bans: Dean Rusby Suggests Universities Ignore Credentials from States with Such Laws," *New York Times*, July 12, 1925.
8. H. L. Mencken, "Sickening Doubts About the Value of Publicity," *Baltimore Evening Sun*, July 9, 1925.
9. Theodore C. Mercer, "A Note on Rhea County," *Tennessee Historical Society* 35 (1976): 92.
10. *James Supply & Hardware Co. v. Dayton Coal & Iron Co.*, 223 F. 991 (6th Cir. 1915).
11. Edward J. Larson, *Summer for the Gods: The Scopes Trial and America's Continuing Debate Over Science and Religion* (New York: Basic Books, 2020), 24.
12. Larson, *Summer for the Gods*, 92–93.
13. *In re Dayton Coal & Iron Co.*, 291 F. 390, 391 (E.D. Tenn. 1922).
14. William Jennings Bryan and Mary Baird Bryan, *The Memoirs of William Jennings Bryan* (Philadelphia: United Publishers of America, 1925), 32–33.
15. Jerry R. Tompkins, "Memoir of a Belated Hero," *American Biology Teacher* 34 (1972): 383.
16. Larson, *Summer for the Gods*; Tompkins, "Memoir of a Belated Hero"; Randy Moore, "Creationism in the United States: I. Banning Evolution from the Classroom," *American Biology Teacher* 60 (1998): 486.
17. Arthur Garfield Hays, *Trial By Prejudice* (New York: Covici-Friede, 1933), 3–22, 326.
18. Andrew E. Kersten, *Clarence Darrow: American Iconoclast* (New York: Hill and Wang, 2011), 238.
19. Arthur Garfield Hays, *Let Freedom Ring* (New York: Boni and Liveright, 1928, 26.
20. Bobby E. Hicks, "The Great Objector: The Public Career of Dr. John R. Neal" (August 1968) (master's thesis available at Trace: Tennessee Research and Creative Exchange); "Lawyer in '25 Tenn. Monkey Trial Dies," *Morristown Gazette Mail*, November 23, 1959; "John Randolph Neal," *Chattanooga Times*, November 24, 1959.

21. "Evolution Is Big Question: Freedom of Education Is on Trial," *Daily Democrat-Forum and Maryville Tribune*, June 4, 1925; Hicks, "The Great Objector."
22. A man named Sue, Hicks would later claim to have been the inspiration of the song "A Boy Named Sue," popularized by Johnny Cash, having once met that song's original author. "Johnny Cash Is Indebted to a Judge Named Sue," *New York Times*, July 12, 1970.
23. Both used significant argument time emphasizing cultural signifiers and denigrating the Northerners on the defense team. For example, McKenzie asked, theatrically, whether "these gentlemen have any laws in the great metropolitan city of New York" or "in the great white city of the northwest that will throw any light" on the case. Transcript, 56–59.
24. Robert C. Cottrell, *Roger Nash Baldwin and the American Civil Liberties Union* (New York: Columbia University Press, 2000), 7–10.
25. Cottrell, *Roger Nash Baldwin*, 156; Stern and DiFore, "Dogging Darwin," 49.
26. Charles Evan Hughes, "Address to the American Bar Association," September 2, 1925, in Merlo J. Pusey, *Charles Evan Hughes*, 2 vols. (New York: Macmillan, 1951), 2:621.
27. Hays, *Let Freedom Ring*, dedication page.
28. Cottrell, *Roger Nash Baldwin*, 155–157.
29. "Shouts of 'Treason' as Malone Speaks," *New York Times*, November 6, 1917; "Hillquit Urges Peace to the End," *New York Times*, November 5, 1917; "Hillquit Gives President Only Divided Loyalty," *New York Times*, October 30, 1917; "Malone for Hillquit; Ex-Collector Indorses Socialist Candidate and His Program for the City," *New York Times*, October 27, 1917.
30. Arthur Garfield Hays, "The Scopes Trial," in Gail Kennedy, ed., *Evolution and Religion: The Conflict Between Science and Theology in Modern America* (Boston: D. C. Heath & Co., 1957), 35–36.
31. *Knoxville Sentinel*, July 12, 1925, in W. C. Curtis, *Scopes Trial Scrapbook* (University of Missouri Archives).
32. Transcript, 47–74.
33. Transcript, 51.
34. Transcript, 53.
35. 262 U.S. at 398.

36. Mark DeWolfe Howe, ed., *Holmes-Laski Letters: The Correspondence of Mr. Justice Holmes and Harold J. Laski, 1916–1935* (Cambridge, MA: Harvard University Press, 1953), 249.
37. *Malone v. Jersey City*, 27 N.J.L. 536, 537 (1859); Ex Parte Fleming, 60 Miss. 910 (1883).
38. *United States v. O'Neil*, 11 F.3d 292, 296 (1st Cir. 1993). (It is "a bit of common sense that has been recognized in virtually every legal code from time immemorial.")
39. Jason S. Thaler, "Public Housing Consent Clauses: Unconstitutional Condition or Constitutional Necessity?," *Fordham Law Review* 63 (1995): 1777, 1797; Peter Westen, "The Rueful Rhetoric of 'Rights,'" *UCLA Law Review* 33 (1986): 977, 986, 1010; Richard A. Epstein, "Unconstitutional Conditions, State Power, and the Limits of Consent," *Harvard Law Review* 102 (1988): 4, 35; *City of Lakewood v. Plain Dealer Publishing Co.*, 486 U.S. 750, 785–86 (1988) (White, J., dissenting); *Bennett v. Metro. Gov't of Nashville & Davidson Cnty*, 977 F.3d 530, 548 (6th Cir. 2020) (Murphy, C. J., concurring); *United States v. Black*, 512 F.2d 864, 869 (9th Cir. 1975); *Blue Cross and Blue Shield of N.C. v. Jemsek Clinic, P.A.*, 506 B.R. 694, 700 (Bankr. W.D.N.C. 2014).
40. *Commonwealth v. Davis*, 39 N.E. 113, 113 (Mass. 1895); *McAuliffe v. City of New Bedford*, 29 N.E. 517, 517 (Mass. 1892); *Rippey v. Texas*, 193 U.S. 504, 509–10 (1904); *City & Cnty. of Denver v. Denver Union Water Co.*, 246 U.S. 178, 196–97 (1918) (Holmes, J., dissenting); *Ferry v. Ramsey*, 277 U.S. 88, 94 (1928); *Frost v. R.R. Comm'n of State of Cal.*, 271 U.S. 583, 602 (1926) (Holmes, J., dissenting). *W. Union Tel. Co. v. Kansas ex rel. Coleman*, 216 U.S. 1, 53 (1910) (Holmes, J., dissenting).
41. *McAuliffe v. City of New Bedford*, 29 N.E. 517, 517 (Mass. 1892).
42. 262 U.S. at 399.
43. Transcript, 87.
44. Transcript, 65–67.
45. See, e.g., *Virden v. Crawford County*, No. 23-cv-2071, 2023 WL 5944154 (W.D. Ark. 2023).
46. "The Baiting of Judge Raulston," *New Republic*, July 29, 1925, 249.
47. Transcript, 115.
48. Transcript, 137.
49. "Evolution Trial: The Effort to Bar Expert Witnesses," *Manchester Guardian*, July 17, 1925, 12.

Chapter Seven: Evolution in the Courtroom

1. Transcript, 102.
2. Transcript, 14.
3. Woodrow Wilson to Winterton Curtis, August 29, 1922, and W. C. Curtis to William Bateson, November 25, 1922, in W. C. Curtis, *Scopes Trial Scrapbook* (University of Missouri Archives).
4. Transcript, 115–116.
5. Haeckel's theory was appealing because it provided a seemingly straightforward and visual connection between individual development and the broader process of evolution, linking embryology directly with evolutionary biology. He used embryological illustrations to show that vertebrate embryos at early stages appeared remarkably similar. However, Haeckel's theory lost influence. The idea that ontogeny directly repeats evolutionary history was refuted by the growing understanding of developmental biology and genetics. One major criticism was that Haeckel's theory suggested a linear, progressive development of species from "primitive" to "advanced" forms, which misrepresented the complexity of evolutionary pathways. Additionally, Haeckel was accused of exaggerating or manipulating some of his drawings to make the similarities between embryos of different species appear more pronounced than they actually were. Confusion during the *Scopes* trial about embryology and evolution is understandable, because at that time, while some aspects of Haeckel's biogenetic law remained influential, its larger framework had been discredited. Stephen Jay Gould, *Ontogeny and Phylogeny* (Cambridge, MA: Belknap Press, 1977); Michael K. Richardson and Gerhard Keuck, "Haeckel's ABC of Evolution and Development," *Biological Reviews* 77 (2002): 495–528.
6. Transcript, 133–143.
7. Transcript, 139
8. Transcript, 148.
9. Transcript, 150–153.
10. Kenneth J. Weiss, "John Wigmore on the Abolition of Partisan Experts," *Journal of the American Academy of Psychiatry and the Law* 43, no. 1 (March, 2015): 21–31.
11. E. Donald Elliott, "The Evolutionary Tradition in Jurisprudence," *Columbia Law Review* 85 (January 1985): 47.
12. John H. Wigmore, "To Abolish the Partisanship of Expert Witnesses, as Illustrated in the Loeb-Leopold Case," *Journal of the*

American Institute of Criminal Law and Criminology 15, no. 3 (November 1924): 341–343.

13. Transcript, 154.
14. Transcript, 202.
15. Transcript, 241–251.
16. Transcript, 234–238.
17. Transcript, 238–241.
18. Transcript, 263–280. See also Roger Lewin, *Bones of Contention: Controversies in the Search for Human Origins* (Chicago: University of Chicago Press, 1987, 2nd ed., 1997). Science writer Lewin's title says it all, as he examines the most noteworthy paleoanthropological controversies in the search for human origins—the Piltdown Man forgery, included—where a few bone fragments make or break academic careers. Ian Tattersall, *The Strange Case of the Rickety Cossack: and Other Cautionary Tales from Human Evolution* (New York: St. Martin's Press, 2015). Paleontologist Tattersall, too, provides a critical look at the hyper-competitive world of paleoanthropology.
19. Curtis would later write letters to Missouri legislators, helping to build support against an unsuccessful attempt to enact a Butler Act analogue in his adopted state. Part of the strategy was for sympathetic legislators to introduce tongue-in-cheek amendments to the bill, thereby making the effort appear ridiculous. One such rider mandated that the law be construed only to prohibit any teaching that "man was once possessed of a prehensile tail." Notes in W. C. Curtis, *Scopes Trial Scrapbook* (University of Missouri Archives).
20. Transcript, 254–263.
21. Transcript, 119–133.
22. Adam R. Shapiro, *Trying Biology: The Scopes Trial, Textbooks, and the Antievolution Movement in American Schools* (Chicago: University of Chicago Press, 2013), 63–65.
23. George William Hunter, *A Civic Biology: Presented in Problems,* 194–95 (1914) in Sheldon Norman Grebstein, ed., *Monkey Trial: The State of Tennessee vs. John Thomas Scopes* (New York: Houghton Mifflin, 1960), 30.
24. William Jennings Bryan, "Who Shall Control?," in William Jennings Bryan and Mary Baird Bryan, *The Memoirs of William Jennings Bryan* (Philadelphia: United Publishers of America, 1925), 535.
25. Shapiro, *Trying Biology*, 62–63.

Chapter Eight: Conviction

1. Gregg Jarrett and Don Yaeger, *The Trial of the Century* (New York: Threshold Editions, 2023) and Brenda Wineapple, *Keeping the Faith: God, Democracy, and the Trial That Riveted a Nation* (New York: Random House, 2024) both resurrect these details.
2. H. L. Mencken, "Fair Trial Beyond Ken," *Baltimore Evening Sun*, July 16, 1925.
3. Lawrence W. Levine, *Defender of the Faith: William Jennings Bryan: The Last Decade, 1915–1925* (Cambridge, MA: Harvard University Press, 1987); Kevin P. Lee, "Inherit the Myth: How William Jennings Bryan's Struggle with Social Darwinism and Legal Formalism Demythologize the Scopes Monkey Trial," *Capitol University Law Review* 33 (2004): 361. This attempted rehabilitation has been only partially successful. Even the Christian philosopher Peter van Inwagen, more sympathetic to Bryan's side than many commentators, opened his discussion of these events by conceding that "Bryan may have been a fool in many respects." Peter van Inwagen. *God, Knowledge, and Mystery: Essays in Philosophical Theology* (Ithaca, NY: Cornell University Press, 1995), 139.
4. Richard Hofstadter, *The American Political Tradition and the Men Who Made It* (New York: Knopf, 1948), 198–202.
5. Transcript, 178.
6. Transcript, 168.
7. With perhaps, surprisingly, the exception of his son, whose narrow and precise arguments about admissibility most closely resembled tactics that would be adopted by a modern government attorney.
8. As noted in chapter 7, Haeckel's ideas persisted in educational materials and public discourse, despite much of the scientific community's having moved beyond them. The ideas, although overly simplistic and no longer fully accurate, provided a visually and conceptually appealing way to try to explain evolutionary theory, which made them useful in a legal setting like the *Scopes* trial, where complex scientific ideas needed to be communicated to a lay audience. Furthermore, many textbooks at the time had not yet updated their content to reflect more current understandings of embryology and evolution, allowing these older ideas to persist in academic circles and public debates. Stephen Jay Gould, *Ontogeny and Phylogeny* (Cambridge, MA: Belknap Press, 1977); Michael K.

Richardson and Gerhard Keuck, "Haeckel's ABC of Evolution and Development," *Biological Reviews* 77 (2002): 495–528.

9. Transcript, 173.
10. Transcript, 292.
11. Robert L. Trivers, "The Evolution of Reciprocal Altruism," *Quarterly Review of Biology* 46, no. 1 (1971): 35–57.
12. William Jennings Bryan and Mary Baird Bryan, *The Memoirs of William Jennings Bryan* (Philadelphia: United Publishers of America, 1925), 544–545; Transcript, 179.
13. Originating in Nietzsche's *The Gay Science* (1882) and *Thus Spoke Zarathustra* (1883–1885).
14. Transcript, 182.
15. William Jennings Bryan, "Last Message of William J. Bryan," accessed November 3, 2024, https://librarycollections.law.umn.edu/documents/darrow/last_message-of_W_J_Bryan.pdf.
16. Charles H. Pence, "Nietzsche's Aesthetic Critique of Darwinism," *History & Philosophy of the Life Sciences* 33 (2011): 165 (cataloging Nietzsche's critiques of Darwinism); Rose Pfeffer, "Eternal Recurrence in Nietzsche's Philosophy," *Review of Metaphysics* 19 (1965): 276, 286–287 ("Nietzsche's own statements express a rejection of Darwin's concept of development.").
17. Bryan and Bryan, *Memoirs*, 548.
18. Defendant's Brief, 2–4.
19. *Scopes v. State*, 278 S.W. 57 (Tenn. 1925).
20. Defendant's Brief, 35.
21. The defense explicitly rejected the contention that "the power to prohibit implies the power to limit," arguing that "this does not follow" and citing to one field where greater-includes-the-lesser argument had been rejected. The reality is somewhat more complicated. Greater-includes-the-lesser argument has deep roots in legal reasoning and is valid in many—but not all—circumstances. It has thus been described as a "trap," as it is often but not always accurate. Michael Herz, "Justice Byron White and the Argument That the Greater Includes the Lesser," *BYU Law Review* 1994: 227.
22. Brief, 66.
23. Brief, 79.
24. Brief, 101.
25. Opposition Brief, 21.

26. Opposition Brief, 12.
27. Opposition Brief, 48.
28. *Scopes v. Tennessee*, 289 S.W. 363 (Tenn. 1927).
29. "Justice Chambliss, of Tennessee, 83," *New York Times*, October 1, 1947, 29.
30. John Thomas Scopes to Clarence Darrow, November 15, 1927, and John Thomas Scopes to Clarence Darrow, February 14, 1927, in *The Clarence Darrow Letters*, https://librarycollections.law.umn.edu/darrow/letters.
31. Jerry R. Tompkins, "Memoir of a Belated Hero," *American Biology Teacher* 34 (1972): 383.

Chapter Nine: Synthesis, Resurrection, and the Shadow of *Scopes*

1. Not to be confused with "Neo-Darwinism," the term used by Darwin's protégé, George John Romanes, in 1905 to describe a version of natural selection that, following Weismann, denied a role for the inheritance of acquired characteristics.
2. There has been considerable research on the Modern Synthesis as a historical event, with different perspectives on the names of the key players and the timing of milestones. The classic account is Ernst Mayr's and William B. Provine's edited volume, *The Evolutionary Synthesis: Perspectives on the Unification of Biology* (Cambridge, MA: Harvard University Press, 1980). Many issues remain contested, however, including "such fundamental questions as to *what, where,* and *when* the Modern Synthesis took place, and even who was involved. Philippe Huneman, "Special Issue Editor's Introduction: 'Revisiting the Modern Synthesis,'" *Journal of the History of Biology* 52 (2019): 509–518. Our core decisions about inclusion here tie to the evolutionary issues that were connected and central to the *Scopes* trial.
3. Genetics, developmental physiology, ecology, systematics, paleontology, cytology, and mathematical biology were all noted by Julian Huxley in his *Evolution: The Modern Synthesis* (London: George Allen & Unwin, 1942).
4. Each of the Synthesists has been the subject of much biographical exploration, and our short and selective summaries here aim to provide the reader a sense of what each of the core contributors brought to the Synthesis, especially in the way it corrected or revised notions about evolution that were expressed at the *Scopes*

trial. Philippe Huneman includes a recent list of such works: Huneman, "Revisiting the Modern Synthesis," 509–518.

5. R. A. Fisher, *Statistical Methods for Research Workers* (Edinburgh: Oliver & Boyd, 1925).
6. R. A. Fisher, *The Genetical Theory of Natural Selection* (Oxford: Clarendon Press, 1930).
7. Leonard Darwin would become the second president of the Eugenics Education Society in 1911. The organization became the Eugenics Society in 1924.
8. Stephen Jay Gould, *The Structure of Evolutionary Theory* (Cambridge, MA: Harvard University Press, 2002), 508–509.
9. William Jennings Bryan, "Last Message of William J. Bryan," accessed July 12, 2024, https://librarycollections.law.umn.edu/documents/darrow/last_message-of_W_J_Bryan.pdf.
10. Fisher's ideas here were later refined by biologist Amotz Zahavi, who explained that sexual selection selects for *costly* traits because only fit individuals will be able to bear the cost of having extravagant traits (such as, again, a peacock's tail). Thus, the extravagant trait ultimately conveys information about overall fitness of potential mates.
11. As a student, Haldane had mastered four languages (in addition to English) and, during his distinguished career as an evolutionary geneticist, also made contributions to chemistry and mathematics.
12. In the first of a series of ten papers he wrote on quantitative approaches to natural selection between 1924 and 1934. Reprinted as J. B. S. Haldane, "A Mathematical Theory of Natural and Artificial Selection—I," *Bulletin of Mathematical Biology* 52 (1990): 209–240.
13. John Burdon Haldane, *The Causes of Evolution* (London: Longmans, Green and Co., 1932), vi, accessed November 3, 2024, https://archive.org/details/causesofevolutio00hald_0/page/n9/mode/2up?q=%22cannot+prevail+against+natural+selection%22+.
14. The process during meiosis—i.e., when haploid gametes are formed from a diploid precursor—of pairs of chromosomes physically lining up and their strands being able to exchange genetic material. Recombination shuffles genes in a random process that creates much of the genetic variation in populations.
15. J. B. S. Haldane, "The Combination of Linkage Values and the Calculation of Distances between the Loci of Linked Factors," *Journal of Genetics* 8, no. 3 (1919): 299–309.

16. J. B. S. Haldane, "The Cost of Natural Selection," *Journal of Genetics* 55, no. 3 (1957): 511–524.
17. Haldane, *The Causes of Evolution*, 131.
18. Wright died at age ninety-eight of complications following a fall on an icy winter's walk near his Madison, Wisconsin, apartment. His *New York Times* obituary (March 4, 1988) noted he had published two hundred scientific papers *after* his retirement from the University of Wisconsin in 1960. His intellect was keen to the very end. One of us (HG), as a Wisconsin zoology graduate student, had the chance to talk with Wright on several occasions and witness his still deeply challenging questions, at nearly ninety years of age, to prominent (and suddenly nervous) visiting speakers in evolutionary biology.
19. The different conceptual emphases by Fisher and Wright were apparently exacerbated by the fact that they did not like each other very much. James F. Crow, "Mid-Century Controversies in Population Genetics," *Annual Review of Genetics* 42 (2008): 1–16.
20. Sewall Wright, "Evolution in Mendelian Populations," *Genetics* 16, no. 2 (1931): 97.
21. Wright's 1932 metaphor of the "adaptive landscape" has had lasting impact on thinking about how species evolve. Sewall Wright, "The Roles of Mutation, Inbreeding, Crossbreeding, and Selection in Evolution," *Proceedings of the 6th International Congress of Genetics* 1 (1932): 356–366. For a sense of the idea, imagine a contour map that shows ground elevation and depression. In an adaptive landscape, greater elevation represents higher fitness ("peaks"), and depressions lower fitness ("valleys"). Points on the map represent the average fitness of individuals in a hypothetical population of different genotypes (or allele frequencies). Evolutionary forces such as natural selection, mutation, and genetic drift, will affect the elevation point on the map representing the mean fitness of the population. Mean fitness can go up or down. While this highly simplified account of the adaptive landscape concept might make it seem straightforward, it has proved challenging and controversial, and to this day continues to be assessed and critiqued by various theorists.
22. The Whitney South Sea Expedition (1920–1941), financed by philanthropist Harry Payne Whitney, served primarily to collect bird specimens for the American Museum of Natural History.

23. "Scopes Trial, 6th & 7th Days," University of Minnesota Libraries, 254, accessed November 3, 2024, https://librarycollections.law.umn.edu/documents/darrow/Scopes%206th%20&%207th%20days.pdf.
24. "Scopes Trial, 6th & 7th Days," University of Minnesota Libraries, 270, accessed November 3, 2024, https://librarycollections.law.umn.edu/documents/darrow/Scopes%206th%20&%207th%20days.pdf.
25. Ernst Mayr, *The Growth of Biological Thought: Diversity, Evolution, and Inheritance* (Cambridge, MA: Harvard University Press, 1982), 562.
26. Allopatric refers to the condition where populations (or species) occupy a geographic region different from that of another population or species (cf. "sympatric," where geographic regions overlap).
27. Of the mathematical Synthesists, Mayr's views on speciation and the importance of small semi-isolated populations align most with Sewall Wright's. Stephen Jay Gould suggests some unspecified friction between Wright and Mayr, at least during the mid-1970s: Mayr refused to invite Wright to the 1974 Workshop on the Evolutionary Synthesis conference that included most of the surviving architects of the Synthesis. Gould, *The Structure of Evolutionary Theory*, 519.
28. Mayr, *The Growth of Biological Thought*, 607.
29. Francisco J. Ayala and Walter M. Fitch, "Genetics and the Origin of Species: An Introduction," *Proceedings of the National Academy of Sciences* 94, no. 15 (1997): 7691–7697.
30. Dobzhansky addressed this shortcoming directly by collaborating with Sewall Wright, who did the mathematical heavy lifting. William B. Provine, *Sewall Wright and Evolutionary Biology* (Chicago: University of Chicago Press, 1986).
31. Theodosius Dobzhansky, "Nothing in Biology Makes Sense Except in the Light of Evolution," *American Biology Teacher* 35, no. 3 (1973): 125–129. A bit of presumption on our part, but perhaps this often-cited claim would be more strictly true had Dobzhansky said, "Nothing in Biology Makes *Complete* Sense. . . ."
32. Lauri Lebo, *The Devil in Dover: An Insider's Story of Dogma v. Darwin in Small Town America* (New York: New Press, 2008), 99.

Chapter Ten: Eugenics, Depression, and the Road to War

1. Bobby E. Hicks, "The Great Objector: The Public Career of Dr. John R. Neal" (August 1968) (master's thesis available at Trace: Tennessee Research and Creative Exchange); "Lawyer in '25

Tenn. Monkey Trial Dies," *Morristown Gazette Mail*, November 23, 1959; "John Randolph Neal," *Chattanooga Times*, November 24, 1959.

2. Melvin I. Urofsky, "The Roosevelt Court," in William Chafe, ed., *The Achievement of American Liberalism: The New Deal Legacy* (New York: Columbia University Press, 2003); Laura Kalman, "Law, Politics, and the New Deal(s)," *Yale Law Journal* 108 (1999): 2165. For a contrary view, see Barry Cushman, "The Secret Lives of the Four Horsemen," *Virginia Law Review* 83 (1997): 559.
3. 300 U.S. 379 (1937).
4. Andrew E. Kersten, *Clarence Darrow: American Iconoclast* (New York: Hill and Wang, 2011), 238–239.
5. William Jennings Bryan and Mary Baird Bryan, *The Memoirs of William Jennings Bryan* (Philadelphia: United Publishers of America, 1925), 548.
6. Paul Weindling, "Julian Huxley and the Continuity of Eugenics in Twentieth-Century Britain," *Journal of Modern European History* 10, no. 4 (2012): 480–499.
7. The Sterilisation Bill of 1931 was introduced by Labour MP and secretary of the Eugenics Society, Major A. G. Church. The bill sought to legalize the voluntary sterilization of "mental defectives," but was voted down by a decided margin, aided by opposition from the Catholic Church. Shortly after, compulsory sterilization laws in Nazi Germany likely limited further attempts to resurrect the bill in Britain.
8. John Burdon Haldane, *The Causes of Evolution* (London: Longmans, Green and Co., 1932), 70, accessed November 3, 2024, https://archive.org/details/causesofevolutio00hald_0/page/n9/mode/2up?q=%22cannot+prevail+against+natural+selection%22+.
9. Samanth Subramanian, *A Dominant Character: The Radical Science and Restless Politics of J. B. S. Haldane* (London: Atlantic Books, 2020). Evolutionary biologist Brian Charlesworth disagrees with this assessment of Haldane's altered view of eugenics in his review of Subramanian's biography, in *The FASEB Journal* 35, no. 11 (November 2021): e21257.
10. Daniel J. Kevles, *In the Name of Eugenics: Genetics and the Uses of Human Heredity* (Berkeley: University of California Press, 1985), 311–319.
11. A. W. Edwards, "The Genetical Theory of Natural Selection," *Genetics* 154, no. 4 (2000): 1419–1426.

12. Although the word "eugenics" has long fallen from fashion, this idea remains present in the popular culture. It is depicted, for example, in the 2006 film *Idiocracy* (Mike Judge, director, Twentieth Century Fox, 2006). For further discussion, see Martin Kolk and Kieron Barclay, "Cognitive Ability and Fertility Among Swedish Men Born 1951–1967: Evidence from Military Conscription Registers," *Proceedings Biological Sciences* 286 (May 2019): 20190359; Susan Currell, "'This May Be the Most Dangerous Thing Donald Trump Believes': Eugenic Populism and the American Body Politic," *Amerikastudien* 64, no. 2 (2019): 291–302.
13. He wrote in 1938: "I do not see that much can be done with the Eugenics Society, as its present directors of policy are strongly entrenched and appear almost impervious to scientific advice."
14. Gould specifically mentions W. B. Provine's classic treatment, *The Origin of Theoretical Population Genetics* (Chicago: University of Chicago Press, 1971).
15. Stephen Jay Gould, *The Structure of Evolutionary Theory* (Cambridge, MA: Harvard University Press, 2002), 512.
16. Richard J. Evans, "R. A. Fisher and the Science of Hatred," *New Statesman*, July 28, 2020.
17. In a similarly motivated decision, Emory University in Atlanta (where HG is a faculty member) removed Robert M. Yerkes' name from what had been the Yerkes National Primate Research Center after a review by the university's Committee on Naming Honors. The origins of the Primate Center date to 1925 (the year of *Scopes*) at Yale University, and the move to Emory was in 1956, the year Yerkes died and at which time the Center took on his name. Yerkes had been a pioneering animal psychologist, the first to promote the potential of nonhuman primate research. The recommended name change was due to Yerkes' past support for, and involvement with, eugenics in the United States. The nomination to remove the name was submitted by a member of the Yerkes family serving on the university's Board of Visitors.
18. Walter Bodmer et al., "The Outstanding Scientist, R. A. Fisher: His Views on Eugenics and Race," *Heredity* 126, no. 4 (2021): 565–576.
19. The authors' conflict of interest statement notes that they are all trustees of the Fisher Memorial Trust. A photo of the Fisher stained-glass window is available at the site's webpage: http://www.senns.uk/FisherWeb.html.

20. *Preterm-Cleveland v. McCloud*, 994 F.3d 512, 587 (6th Cir. 2021) (Donald, dissenting).
21. *Box v. Planned Parenthood of Indiana & Kentucky*, 587 U.S. 490, 495 (2019) (Thomas, J., concurring).
22. Nathan Frankowski, director, *Expelled: No Intelligence Allowed.* Vivendi Entertainment and Rocky Mountain Films, 2008.
23. Paul A. Lombardo, "Disability, Eugenics, and the Culture Wars," *St. Louis University Journal of Health Law and Policy* 2 (2008): 57, 70.
24. *Colbert v. Infinity Broad. Corp.*, 423 F. Supp. 2d 575, 585–86 (N.D. Tex. 2005).
25. Edward Lurie, "Louis Agassiz and the Races of Man," *Isis* 45, no. 3 (1954): 227–242.

Chapter Eleven: The Midcentury Moment

1. William L. Shirer, *The Collapse of the Third Republic: An Inquiry into the Fall of France* (New York: Simon & Schuster 1968), 138. Shirer is perhaps best remembered for his history of the Nazi movement, informed by his experience as a foreign correspondent stationed in Germany at the beginning of the war, which he recounted in *The Rise and Fall of the Third Reich* (Stratford-upon-Avon: Arrow, 1991).
2. "Five Arrested in Roundup of War Weapons," *Los Angeles Times*, March 2, 1947.
3. Arthur Corbin, "The Law and the Judges," *Yale Review* 3 (1914): 238.
4. E. Donald Elliott, "The Evolutionary Tradition in Jurisprudence," *Columbia Law Review* 85 (January 1985): 46–50.
5. Elliott, "The Evolutionary Tradition in Jurisprudence," 59.
6. Nancy Tucker, *Patterns in the Dust* (New York: Columbia University Press, 1983).
7. *Harisiades v. Shaughnessy*, 342 U.S. 580 (1952).
8. *Joint Anti-Fascist Refugee Comm. v. McGrath*, 341 U.S. 123, 174–182 (1951) (Douglas, J., concurring).
9. Ellen Schrecker, "Archival Sources for the Study of McCarthyism," *Journal of American History* 75, no. 1 (1988): 197–208; "Rabbi Schultz Backs Gitlow's Testimony," *New York Times*, September 16, 1953.
10. Gad Guterman, "Field Tripping: The Power of 'Inherit the Wind,'" *Theatre Journal* 60 (2008): 563; Albert Wertheim, "The McCarthy Era and the American Theatre," *Theatre Journal* 34 (1982): 211.
11. Guterman, "Field Tripping."

12. Daniel J. Kevles, *In the Name of Eugenics: Genetics and the Uses of Human Heredity* (Berkeley: University of California Press, 1985). Some historians point to the publication of Kevles's well-received book, itself, as a turning point. The book's remarkable success with particular audiences aligns with the popularity of its message at a time when postmodernist biophobia, and also some well-deserved scientific skepticism, was being directed at the emerging fields of sociobiology and evolutionary psychology with their emphasis on the inheritance of behavior.
13. *Skinner v. Oklahoma*, 316 U.S. 535, 541 (1942).
14. Julian Huxley, "Eugenics in Evolutionary Perspective," in *Evolutionary Humanism* (Buffalo, NY: Prometheus Books, 1992).
15. Huxley, *Evolutionary Humanism*, 271.
16. Francisco José Ayala and James W. Valentine, *Evolving: The Theory and Processes of Organic Evolution* (Boston: Addison Wesley, 1979), 407–413.
17. The negative connotations that the word had assumed by the 1960s will be clear to many *Star Trek* fans, who may have first heard of eugenics through the character Khan, appearing in the original series episode "Space Seed" and the later movie, *Star Trek II: The Wrath of Khan*. Khan was the product of a genetic engineering project gone wrong during Earth's past. In "Space Seed," Captain Kirk and the crew discover a ship holding some of these genetically modified and enhanced humans, including Khan, preserved in suspended animation. These antagonists possessed superhuman physical strength and intelligence but were tyrannical and ambitious, conquering much of the planet during so-called Eugenics Wars of the 1990s.
18. Fay-Cooper Cole, "A Witness at the Scopes Trial," *Scientific American* 200, no. 1 (January 1959): 120–131.
19. Harry Kalven Jr., "A Commemorative Case Note: *Scopes v. State*," *University of Chicago Law Review* 27 (1960): 505; Thomas I. Emerson and David Haber, "The Scopes Case in Modern Dress," *University of Chicago Law Review* 27 (1960): 522; Malcolm P. Sharp, "Science, Religion, and the Scopes Case," *University of Chicago Law Review* 27 (1960): 529.
20. Emerson and Haber, "The Scopes Case in Modern Dress."
21. Yanek Mieczkowski, *Eisenhower's Sputnik Moment: The Race for Space and World Prestige* (Ithaca, NY: Cornell University Press: 2013).

22. Gerald Skoog, "Topic of Evolution in Secondary School Biology Textbooks: 1900–1977," *Science Education* 63 (1979): 621; *McLean v. Arkansas Bd. of Ed.*, 529 F. Supp. 1255, 1259 (E.D. Ark. 1982).
23. See Ronald Ladouceur, "Ella Thea Smith and the Lost History of American High School Biology Textbooks," *Journal of the History of Biology* 41 (2008): 435.
24. John L. Rudolph, "Teaching Materials and the Fate of Dynamic Biology in American Classrooms After Sputnik," *Technology & Culture* 53 (2012); Ruth C. Stern and J. Herbie DiFonzo, "Dogging Darwin: America's Revolt Against the Teaching of Evolution," *Northern Illinois University Law Review* 36 (2016); *McLean v. Arkansas Bd. of Ed.*, 529 F. Supp. 1255, 1259 (E.D. Ark. 1982).
25. Arnold B. Grobman, "Editorial: Issues in Science Education," *Science Teacher* 35 (1968): 18; Cal Ledbetter Jr., "The Antievolution Law: Church and State in Arkansas," *Arkansas Historical Quarterly* 38 (1979): 299.
26. "Anti-Evolution Upheld," *Science News* 91 (1967): 569.
27. "Anti-Evolution Upheld."
28. Ledbetter, "The Antievolution Law."
29. Emerson and Haber, "The Scopes Case in Modern Dress."
30. Randy Moore, "Thanking Susan Epperson," *American Biology Teacher* 60 (1998): 642 ("the Gazette wondered, tongue-in-cheek, if perhaps Mr. Bennett had been referring to the Green Bay Packers' linebacker, Ray Nitsky" [sic]).
31. *State v. Epperson*, 416 S.W.2d 322, 322 (Ark. 1967), *rev'd sub nom. Epperson v. Arkansas*, 393 U.S. 97 (1968).
32. *Epperson*, 393 U.S. at 114–15 (Harlan, J., concurring).
33. See Fred P. Graham, "Darwin and That Theory Are Back in Court," *New York Times*, October 17, 1968; John P. MacKenzie, "Supreme Court Justices Wrestling with Pesky Arkansas 'Monkey Law,'" *Washington Post*, October 20, 1968 (stating that Epperson was "now a housewife in Oxon Hill, MD").
34. Another plaintiff, a parent of an Arkansas schoolchild, had intervened in the case alongside Epperson, but the parent-intervenor's case was not without its own standing problems. *Epperson*, 393 U.S. at 110 (Black, J., concurring).
35. Milton V. Freeman, "Abe Fortas: A Man of Courage," *Yale Law Journal* 91 (1982): 1052, 1055–1060.

36. Justin Crowe and Christopher Karpowitz, "Where Have You Gone, Sherman Minton? The Decline of the Short-Term Supreme Court Justice," *Perspectives on Politics* 5 (2007): 425; Geoffrey Stone, "Understanding Supreme Court Confirmations," *Supreme Court Review* 2010 (2011): 381.
37. 372 U.S. 726, 732 (1963).
38. William Domnarski, *The Great Justices, 1941–54* (Ann Arbor: University of Michigan Press 2006): 99–128.
39. E.g., Wallace Mendelson, "Hugo Black and Judicial Discretion," *Political Science Quarterly* 85 (1970): 17.
40. Richard A. Posner, "In Memoriam: William J. Brennan, Jr.," *Harvard Law Review* 111 (1997): 9–10.
41. *Epperson v. Arkansas*, 393 U.S. 97, 114 (1968) (Black, J., concurring).
42. Adrian Vermeule, "Should We Have Lay Justices?," *Stanford Law Review* 59 (April, 2007): 1750–1753.
43. Founded in 1892 by Isaac Jones Wistar, prominent Philadelphia lawyer and vice-president of the Pennsylvania Railroad, as a nonprofit institution to focus on biomedical research and training.
44. Roger Lewin, "Biology Is Not Postage Stamp Collecting: Ernst Mayr, the Eminent Harvard Evolutionist, Explains Why He Thinks Some Physical Scientists Have a Problem with Evolution," *Science* 216, no. 4547 (1982): 718–720.

Chapter Twelve: *Lemon* and Peppered Moths

1. *Smith v. State*, 242 So. 2d 692, 698 (Miss. 1970).
2. *Lemon v. Kurtzman*, 403 U.S. 602 (1971).
3. See *Edwards v. Aguillard*, 482 U.S. 578 (1987); *Selman v. Cobb Cnty. Sch. Dist.*, 449 F.3d 1320 (11th Cir. 2006); *Freiler v. Tangipahoa Par. Bd. of Educ.*, 185 F.3d 337 (5th Cir. 1999); *Peloza v. Capistrano Unified Sch. Dist.*, 37 F.3d 517 (9th Cir. 1994); *Daniel v. Waters*, 515 F.2d 485 (6th Cir. 1975); *Wright v. Houston Indep. Sch. Dist.*, 486 F.2d 137 (5th Cir. 1973); *Kitzmiller v. Dover Area Sch. Dist.*, 400 F. Supp. 2d 707 (M.D. Pa. 2005); *McLean v. Arkansas Bd. of Ed.*, 529 F. Supp. 1255 (E.D. Ark. 1982); *Moeller v. Schrenko*, 554 S.E.2d 198 (Ga. Ct. App. 2001). Cf. *Steele v. Waters*, 527 S.W.2d 72 (Tenn. 1975) (striking down statute banning the teaching of "all occult or satanical beliefs of human origin").

4. See Tenn. Op. Att'y Gen. No. 79–136 (Mar. 26, 1979); 67 Md. Op. Att'y Gen. 26 (1982); Tenn. Op. Att'y Gen. No. 96–025 (Feb. 26, 1996).
5. E.g., *Valent v. New Jersey State Bd. of Educ.*, 274 A.2d 832, 840 (N.J. Ch. Div. 1971). ("The disputed area of evolution, still disputed after all these years, is a matter of one belief in a scientific fact which does not intrude as long as other [sic] doctrine of genesis is given to the children.")
6. *Pickering v. Bd. of Ed. of Twp. High Sch.*, 391 U.S. 563 (1968).
7. *Tinker v. Des Moines Indep. Community Sch. Dist.*, 391 U.S. 563 (1969).
8. By influential evolutionary biologist George C. Williams.
9. G. C. Williams, *Adaptation and Natural Selection: A Critique of Some Current Evolutionary Thought* (Princeton, NJ: Princeton University Press, 1966).
10. John A. Endler, *Natural Selection in the Wild* (Princeton, NJ: Princeton University Press, 1986).
11. Henry Bernard Davies Kettlewell, "Selection Experiments on Industrial Melanism in the Lepidoptera," *Heredity* 9, no. 3 (1955): 323–342; Henry Bernard Davies Kettlewell, "Further Selection Experiments on Industrial Melanism in the Lepidoptera," *Heredity* 10, no. 3 (1956): 287–301.
12. Bryan Campbell Clarke, "Edmund Brisco Ford. 23 April 1901–2 January 1988," *Biographical Memoirs of Fellows of the Royal Society* (London: Royal Society, 1995): 146–168.
13. Laurence M. Cook and John R. G. Turner, "Fifty Per Cent and All That: What Haldane Actually Said," *Biological Journal of the Linnean Society* 129, no. 3 (2020): 765–771.
14. David W. Rudge, "Myths About Moths: A Study in Contrasts," *Endeavour* 30, no. 1 (2006): 19–23.
15. In particular, H. B. D. Kettlewell, "Darwin's Missing Evidence," *Scientific American* 200: (1959): 48–53.
16. Judith Hooper, *Of Moths and Men: An Evolutionary Tale: The Untold Story of Science and the Peppered Moth* (New York: W. W. Norton, 2002).
17. David W. S. Rudge, "Review of *Of Moths and Men: An Evolutionary Tale: Intrigue, Tragedy and the Peppered Moth*, by J. Hooper," *Journal of the History of Biology* 36, no. 1 (2003): 207–209.
18. For example, there is evidence that the actual resting locations of these moths in nature are the undersides of branches, on trunks in

shaded positions, or just below major branch joints. Kettlewell had instead placed his moths in exposed positions on tree trunks. But a replication study that compared the different resting locations found survival patterns that mirrored Kettlewell's original results. R. J. Howlett and M. E. N. Majerus, "The Understanding of Industrial Melanism in the Peppered Moth (*Biston betularia*) (Lepidoptera: Geometridae)," *Biological Journal of the Linnean Society* 30, no. 1 (1987): 31–44. Additional confirmation comes from experimental studies that clearly demonstrate the effect of insect crypsis—camouflage or protective appearance—on bird prey detection. A. T. Pietrewicz and A. C. Kamil, "Visual Detection of Cryptic Prey by Blue Jays (*Cyanocitta cristata*)," *Science* 195, no. 4278 (1977): 580–582.

19. Reviewed in Rudge, "Myths about Moths," 19–23; Cook and Turner, "Fifty Per Cent and All That," 765–771.
20. Jerry A. Coyne, *Why Evolution Is True* (Oxford: Oxford University Press, 2010).
21. Jerry A. Coyne, "Not Black and White," *Nature* 396 (1998): 35–36.
22. Coyne provides a fascinating account of the arc of his views on the peppered moth and Kettlewell on his website: https://why evolutionistrue.com/2012/02/10/the-peppered-moth-story-is -solid/.
23. Stephen Jay Gould, *The Structure of Evolutionary Theory* (Cambridge, MA: Belknap Press, 2002), 518.
24. Historian Peter J. Bowler considers Gould one of the "opponents of ultra-Darwinism," a group who challenged various elements of the Modern Synthesis, and notes some evolutionary biologists saw him as nothing more than a self-promoter and science popularizer. Gould received a lot of attention due to his widely read monthly column in *Natural History* magazine. Peter J. Bowler, *Evolution: The History of an Idea* (Los Angeles: University of California Press, 1989), 362.
25. Dobzhansky's student, Richard Lewontin, was among those who first applied techniques from molecular biology, such as gel electrophoresis, to questions concerning genetic variation in evolution, giving birth to the field of molecular evolution. Lewontin would later be prominent in public controversies both within evolution (over the evidentiary basis for whether traits qualified as adaptations or not, and in the sociobiology debates over human behavior), and

politically (resigning from the National Academy of Sciences in 1971 in protest of secret military research being conducted by the National Research Council in connection with the Vietnam War).

26. Gould, *The Structure of Evolutionary Theory*, 521.
27. Motoo Kimura, "Evolutionary Rate at the Molecular Level," *Nature* 217, no. 5129 (1968): 624–626; Motoo Kimura, *The Neutral Theory of Molecular Evolution* (Cambridge: Cambridge University Press, 1983).
28. Although genetic mutations are random events, they occur at a reasonably constant rate. Not all genes mutate at the same rate, however. The number of differences between comparable gene sequences across two species would increase over time. Thus, the number of mutations in DNA sequences can be used as a measure of time, providing valuable phylogenetic information.
29. Jeffrey D. Jensen et al., "The Importance of the Neutral Theory in 1968 and 50 Years On: A Response to Kern and Hahn 2018," *Evolution* 73, no. 1 (2019): 111–114.
30. Stephen Jay Gould, "Sociobiology: The Art of Storytelling," *New Scientist* 80, no. 1129 (1978): 530–533.
31. Peter Grant taught H. G. his first undergraduate Behavioral Ecology course at McGill University.
32. The Grants and their colleagues focused on the medium ground finches (*Geospiza fortis*) on the island of Daphne Major.
33. Endler, *Natural Selection in the Wild*.
34. He provided a critical review of the various methods in use—some deemed more compelling and complete than others—for demonstrating natural selection and, importantly, discussed the kinds of statistical approaches that he thought were most appropriate for analyzing the relevant data. Endler presents natural selection as a syllogism based on Darwinian reasoning:

    > given three conditions for natural selection (a, variation; b, fitness differences; c, inheritance), two conclusions necessarily follow: (1) differences in trait distributions among age classes or life history stages; and (2) if the population is not at equilibrium, a predictable difference among generations (Endler, *Natural Selection in the Wild*, 26).

    With the criteria he established, Endler identified demonstrations of natural selection in natural populations for over one hundred species of plants and animals.

35. K. P. Harden, "Genetic Determinism, Essentialism and Reductionism: Semantic Clarity for Contested Science," *Nature Review Genetics* 24 (2023): 197–204. Harden's analysis of the terms "genetic determinism," "genetic essentialism," and "genetic reductionism," and the controversies surrounding their usage, are especially useful. In modern usage, the term "genetic determinism" generally has negative connotations in the fields of psychology, behavioral genetics, and evolutionary biology. No researcher wants to be labeled a genetic determinist because it is an accusation that conclusions or contentions about human traits or behaviors have overstated the contribution of genes in explaining variability. See, for example, reviews centering on genetic determinism in Robert Plomin's book *Blueprint: How DNA Makes Us Who We Are* (Cambridge, MA: MIT Press, 2019), such as N. Comfort, "Genetic Determinism Redux," *Nature* 561 (2018): 27; J. Joseph, "A Blueprint for Genetic Determinism," *American Journal of Psychology* 135, no. 4 (2022): 442–454.

Chapter Thirteen: A Punctured Synthesis

1. *Wright v. Houston Indep. Sch. Dist.*, 486 F.2d 137, 138 (5th Cir. 1973).
2. *Daniel v. Waters*, 515 F.2d 485 (6th Cir. 1975).
3. *Steele v. Waters*, 527 S.W.2d 72 (Tenn. 1975).
4. *Willoughby v. Stever*, No. 15574–75 (D.D.C. May 18, 1973); aff'd. 504 F.2d 271 (D.C. Cir. 1974), cert. denied, 420 U.S. 927 (1975).
5. Stephen Jay Gould and Niles Eldredge, "Punctuated Equilibria: An Alternative to Phyletic Gradualism," *Models in Paleobiology* (1972): 82–115.
6. This account is different from the saltationist view that intermediate stages never existed and instead mutations result in individuals that differ immediately and clearly from their ancestors.
7. Duane T. Gish, *Evolution: The Fossils Still Say No!* (Dallas: Institute for Creation Research, 1995). Anti-evolutionist Gish, who quotes Eldredge and Gould, focuses his argument almost exclusively on these gaps in the fossil record, concluding if the record does not support the idea of gradualism, then evolution must not have occurred at all.
8. Kieran Schlegel-O'Brien, "Stephen Jay Gould, From Evolution to Revolution," *Advanced Science News*, August 1, 2022, https://www.advancedsciencenews.com/stephen-jay-gould-from-evolution-to-revolution.

9. G. G. Simpson, *Tempo and Mode in Evolution* (New York: Columbia University Press, 1944). Simpson recognized the sudden appearance of some higher taxa, including some mammalian orders, and refers to such occurrences as "quantum evolution." Gould and Eldredge were not so much focused on the emergence of higher taxa as they were newly emergent closely related species.
10. Stephen Jay Gould, *The Structure of Evolutionary Theory* (Cambridge, MA: Harvard University Press, 2002), 76–77.
11. B. M. Kozo-Polyansky, *Symbiogenesis: A New Principle of Evolution* (Cambridge, MA: Harvard University Press, 2010).
12. Lynn Sagan, "On the Origin of Mitosing Cells," *Journal of Theoretical Biology* 14, no. 3 (1967): 225–232; Lynn Margulis, "The Origin of Plant and Animal Cells: The Serial Symbiosis View of the Origin of Higher Cells Suggests That the Customary Division of Living Things into Two Kingdoms Should Be Reconsidered," *American Scientist* 59, no. 2 (1971): 230–235. (Margulis had been married to Carl Sagan, the well-known astronomer and science communicator.)
13. One of the more perplexing absences in Stephen Jay Gould's extensive writings on evolution is his lack of discussion on Lynn Margulis's theory of symbiogenesis. Given that Gould's own theory of punctuated equilibrium challenged the gradualist views of the Modern Synthesis, one might expect him to engage more deeply with Margulis's work, which also emphasized non-gradual evolutionary changes. Margulis's ideas could have been seen as complementary to Gould's, especially in their shared rejection of strict gradualism. The fact that Gould rarely, if ever, discussed Margulis's work in his popular essays may reflect a broader reluctance within the evolutionary biology community to fully integrate symbiogenesis into the established framework of evolutionary theory at the time. Alternatively, Gould's focus on macroevolutionary patterns and the fossil record may have led him to prioritize other topics over the cellular and microbiological processes emphasized by Margulis. The absence of Margulis's ideas in Gould's writing remains a notable omission, particularly given the potential synergies between their respective challenges to the Modern Synthesis.
14. Dick Teresi, "Discover Interview: Lynn Margulis Says She's Not Controversial, She's Right," *Discover Magazine*, June 16, 2011,

https://www.discovermagazine.com/the-sciences/discover-interview-lynn-margulis-says-shes-not-controversial-shes-right.

15. James A. Lake, "Lynn Margulis (1938–2011)," *Nature* 480, no. 7378 (2011): 458. Her revolutionary paper was submitted to a dozen journals before it was published in the *Journal of Theoretical Biology*.
16. "Lynn Margulis," *The Telegraph*, December 13, 2011.
17. Teresi, "Discover Interview: Lynn Margulis Says She's Not Controversial."
18. *Willoughby v. Stever*, No. 15574–75 (D.D.C. May 18, 1973); aff'd. 504 F.2d 271 (D.C. Cir. 1974), cert. denied, 420 U.S. 927 (1975).
19. Teresi, "Discover Interview: Lynn Margulis Says She's Not Controversial."
20. "Lynn Margulis," *The Telegraph*, December 13, 2011. Phrenology was a pseudoscientific theory that emerged in the late eighteenth and early nineteenth centuries, that posited the shape and size of various regions of the skull could determine a person's character traits, mental abilities, and personality.
21. Teresi, "Discover Interview: Lynn Margulis Says She's Not Controversial."
22. Teresi, "Discover Interview: Lynn Margulis Says She's Not Controversial."
23. Lynn Margulis, *Symbiosis in Cell Evolution: Life and Its Environment on the Early Earth* (New York: W. H. Freeman, 1981); Lynn Margulis, *Symbiotic Planet: A New Look at Evolution* (New York: Basic Books, 1999); Lynn Margulis and Dorion Sagan, *Acquiring Genomes: A Theory of the Origins of Species* (New York: Basic Books, 2002).
24. William Jennings Bryan, "Last Message of William J. Bryan," accessed November 3, 2024, https://librarycollections.law.umn.edu/documents/darrow/last_message-of_W_J_Bryan.pdf.
25. Edward O. Wilson, *Sociobiology. The New Synthesis* (Cambridge, MA: Belknap Press, 1975). Wilson's book, *On Human Nature* (Harvard University Press, 1978), would win a Pulitzer Prize and bring additional controversy because of its contentions about human behavior.
26. Ullica Segerstråle, *Defenders of the Truth: The Battle for Science in the Sociobiology Debate and Beyond* (Oxford: Oxford University Press, 2000). Segerstråle provides a very thorough history of the field during this period.
27. "Genes über Alles," *Time Magazine* 108, no. 24 (December 13, 1976).

28. Wilson, *Sociobiology*, 4, 362.
29. M. D. Sahlins, *The Use and Abuse of Biology: An Anthropological Critique of Sociobiology* (Ann Arbor: University of Michigan Press, 1976), xii. Of note, Sahlins was a political activist, the creator of the campus "teach-in" protesting of the Vietnam War era. Like Richard Lewontin, Sahlins resigned from the National Academy of Sciences in 2013, partially for its support of military research.
30. Transcript, 292.
31. Simpson's review did not stand alone in its critical assessment, and Sahlins himself, acknowledged that his book had been "often rubbished," but maintained his convictions about sociobiology, and especially some of its claims and practitioners. Marshall Sahlins, "The National Academy of Sciences: Goodbye to All That," *Anthropology Today* 29 no. 2 (2013): 1–2.
32. "Genes über Alles."
33. Robert Cooke, "Protesters Douse Harvard Speaker," *Boston Globe*, February 16, 1978, 14.
34. An organization founded in 1973. It disbanded in 1996.
35. Sahlins, *The Use and Abuse of Biology*; Arthur L. Caplan, *The Sociobiology Debate: Readings on Ethical and Scientific Issues* (New York: Harper & Row, 1978); Steven Rose, *Against Biological Determinism* (New York: Viking Press, 1980); Stephen Jay Gould, *The Mismeasure of Man* (New York: W. W. Norton & Company, 1981); Richard C. Lewontin, Steven Rose, and Leon J. Kamin, *Not in Our Genes: Biology, Ideology, and Human Nature* (New York: Pantheon Books, 1984); Stephen R. L. Clark, *Sociobiology: Sense or Nonsense?* (New York: D. Reidel, 1987); Mary Maxwell, ed, *The Sociobiological Imagination* (Buffalo: State University of New York Press, 1991).
36. Richard A. Epstein, "A Taste for Privacy? Evolution and the Emergence of a Naturalistic Ethic," *Journal of Legal Studies* 9, no. 4 (1980): 669–670.
37. William H. Rodgers Jr., "Bringing People Back: Toward A Comprehensive Theory of Taking in Natural Resources Law," *Ecology Law Quarterly* 10, no. 2 (1982): 205–206.
38. John Alcock, *The Triumph of Sociobiology* (Oxford: Oxford University Press, 2001).

Chapter Fourteen: Crusades Begin

1. *Crowley v. Smithsonian Inst.*, 636 F.2d 738 (D.C. Cir. 1980).
2. Randy Moore, "Creationism in the United States V: The McLean Decision Destroys the Credibility of 'Creation Science,'" *American Biology Teacher* 61 (1999): 92.
3. Steven Jay Gould, "A Visit to Dayton," *Natural History* 90, no. 10 (1981): 8–22.
4. Myrna Perez, "Stephen Jay Gould and *McLean v. Arkansas*: Scientific Expertise and the Nature of Science in American Culture 1980–1985," *Between Scientists & Citizens: Proceedings of a Conference at Iowa State University*, June 1–2, 2012 (GPSSA, 2012).
5. Moore, "The McLean Decision," 92.
6. *United States v. O'Brien*, 391 U.S. 367, 384 (1968).
7. *McLean v. Arkansas Bd. of Ed.*, 529 F. Supp. 1255, 1256 (E.D. Ark. 1982).
8. *Keith v. Louisiana Dep't of Educ.*, 553 F. Supp. 295 (M.D. La. 1982).
9. *Aguillard v. Treen*, 440 So. 2d 704 (La. 1983).
10. *Aguillard v. Treen*, 634 F. Supp. 426 (E.D. La. 1985).
11. In 1980, for example, Reagan contended that evolution "is not yet believed in the scientific community to be as infallible as it once was believed." Ruth C. Stern and J. Herbie DiFonzo, "Dogging Darwin: America's Revolt Against the Teaching of Evolution," *Northern Illinois University Law Review* 36 (2016): 36.
12. *Aguillard v. Edwards*, 765 F.2d 1251 (5th Cir. 1985).
13. *Aguillard v. Edwards*, 778 F.2d 225 (5th Cir. 1985).
14. Michael Brant Shermer, "Science Defended, Science Defined: The Louisiana Creationism Case," *Science, Technology, and Human Values* 16 (1991): 517–538; "Amicus Curiae Brief of 72 Nobel Laureates, 17 State Academies of Science, and 7 Other Scientific Organizations," *Aguillard v. Edwards*; Randy Moore and Don Aguillard, "The Courage & Convictions of Don Aguillard," *American Biology Teacher* 61 (1999): 166.
15. In a recent controversy analogous to those concerning R. A. Fisher's support for eugenics, several honorary titles were stripped from Watson for what *Smithsonian Magazine* termed a "decades-long pattern of racist remarks." The complicated legacies of many pioneering scientists of the early twentieth century were by no means restricted to the Synthesists. See "DNA Pioneer James

Watson Loses Honorary Titles over Racist Comments," *Smithsonian Magazine*, January 15, 2019.

16. Shermer, "Science Defended, Science Defined," 517–538.
17. Shermer, "Science Defended, Science Defined," 517–538.
18. Shermer, "Science Defended, Science Defined," 517–538.
19. Shermer, "Science Defended, Science Defined," 517–538.
20. Brief of Nobel Laureates, *Aguillard v. Edwards.*
21. On Brennan's legacy, see Laurence H. Tribe, "In Memoriam: William J. Brennan, Jr.," *Harvard Law Review* 111 (1997): 41, 42. On his jurisprudential differences compared with progressives of an earlier era, see David E. Marion, "Justice William J. Brennan and the Spirit of Modernity," *Polity* 27 (1995): 405.
22. *Edwards v. Aguillard*, 482 U.S. 578 (1987).
23. *Sch. Dist. of Abington Twp. v. Schempp*, 374 U.S. 203, 253 (1963) (Brennan, J., concurring).
24. 482 U.S. at 594.
25. Joelle Anne Moreno, "Extralegal Supreme Court Policy-Making," *William & Mary Bill of Rights Journal* 24, no. 2 (2015): 451–520.
26. "Teach Science as Science: A Teacher's Lawsuit Revisits the Scopes trial," *Los Angeles Times*, July 27, 1994.
27. *Webster v. New Lenox Sch. Dist. No. 122*, 917 F.2d 1004 (7th Cir. 1990).
28. *Peloza v. Capistrano Unified Sch. Dist.*, 782 F. Supp. 1412, 1415 (C.D. Cal. 1992).
29. *Peloza v. Capistrano Unified Sch. Dist.*, 37 F.3d 517 (9th Cir. 1994).
30. *Mozert v. Hawkins Cnty. Bd. of Educ.*, 827 F.2d 1058 (6th Cir. 1987).
31. Jacquelyn Cameron and Randy Moore, "Exploring the Scopes 'Monkey' Trial in Dayton, Tennessee: A Guide to People & Places," *American Biology Teacher* 77 (2015): 333.
32. *Doe v. Porter*, 188 F. Supp. 2d 904 (E.D. Tenn. 2002), *aff'd* 370 F.3d 558 (6th Cir. 2004).
33. *Doe v. Porter*, 188 F. Supp. 2d 904 (E.D. Tenn. 2002).
34. *Doe v. Porter*, 370 F.3d 558 (6th Cir. 2004).

Chapter Fifteen: Backlash and Unraveling

1. Michael J. Klarman, "Brown and Lawrence (and Goodridge)," *Michigan Law Review* 104 (2005): 431, 473.
2. Bork, for example, is generally credited with staking out the originalist position in his 1971 article, "Neutral Principles and

Some First Amendment Problems," *Indiana Law Journal* 47, no. 1 (1971).
3. Joelle Anne Moreno, "Extralegal Supreme Court Policy-Making," *William & Mary Bill of Rights Journal* 24, no. 2 (2015): 451–520.
4. For a discussion of the role of Meese and Bork, see Conor Casey and Adrian Vermeule, "Myths of Common Good Constitutionalism," *Harvard Journal of Law & Public Policy* 45 (2022): 103, 130–131.
5. Justice Scalia, for example, declared that he did not "care if the framers of the Constitution had some secret meaning in mind when they adopted its words." Joe Carter, "Justice Scalia's Two Most Essential Speeches," The Ethics and Religious Liberty Commission of the Southern Baptist Convention, February 18, 2016.
6. Lawrence B. Solum, "Originalism versus Living Constitutionalism: The Conceptual Structure of the Great Debate," *Northwestern University Law Review* 113 (2019): 1243, 1245; Lawrence B. Solum, "The Fixation Thesis: The Role of Historical Fact in Original Meaning," *Notre Dame Law Review* 91 (2015): 1.
7. Amy Coney Barrett, "Substantive Canons and Faithful Agency," *Boston University Law Review* 90 (2010): 109–182.
8. David M. Driesen, "Purposeless Construction," *Wake Forest Law Review* 48 (2013): 97.
9. Barrett, "Substantive Canons and Faithful Agency."
10. *Bd. of Cnty. Comm'rs v. Umbehr*, 518 U.S. 668, 688–89 (1996) (Scalia, J., dissenting).
11. Jamal Greene, "The Age of Scalia," *Harvard Law Review* 130 (2016): 144.
12. John F. Manning, "The New Purposivism," *Supreme Court Review* 2011 (2011): 113.
13. Scott Dodson, "A Darwinist View of the Living Constitution," *Vanderbilt Law Review* 61 (Oct. 2008): 1319.
14. *West Virginia v. EPA*, 142 S. Ct. 2587, 2641 (2022) (Kagan, J., dissenting).
15. Moreno, "Extralegal Supreme Court Policy-Making," 451–520.
16. Sean M. Kammer, "'Whether or Not Special Expertise Is Needed': Anti-Intellectualism, the Supreme Court, and the Legitimacy of Law," *South Dakota Law Review* 63 (2018): 287, 289.
17. 573 U.S. 682.
18. Jeff Tollefson, "The Supreme Court's War on Science," *Nature* 609 (September 15, 2022): 460–462.

19. "Brief of Climate Scientists Michael Oppenheimer, Noah Diffenbaugh, Christopher Field, Stephen Pacala, Daniel Schrag, and Susan Solomon as Amici Curiae in Support of Respondents," *West Virginia v. EPA*.
20. *Loper Bright Enterprises v. Raimondo*, 603 U.S. __ (2024).
21. *Aguillard*, 482 U.S. at 583 n.4.
22. *Aguillard*, at 629 (Scalia, J., dissenting).
23. Stephen Jay Gould, "Justice Scalia's Misunderstanding." *Constitutional Commentary* (1988): 1017.
24. Douglas J. Futuyma, *Science on Trial: The Case for Evolution* (New York: Pantheon Books, 1983).
25. With the research and writings of psychologist Leda Cosmides, and the anthropologists, John Tooby, Donald Symons, and Jerome Barkow. Significant attention came following their publication, in 1992, of *The Adapted Mind: Evolutionary Psychology and the Generation of Culture* (New York: Oxford University Press, 1992).
26. Donald Symons, *The Evolution of Human Sexuality* (New York: Oxford University Press, 1979).
27. Desmond Morris, *The Naked Ape: A Zoologist's Study of the Human Animal* (New York: Jonathan Cape, 1967).
28. Sarah Blaffer Hrdy, "The Evolution of Human Sexuality: The Latest Word and the Last," *Quarterly Review of Biology* (1979): 309–314. Even a review by Hrdy, an anthropologist aligned with sociobiological reasoning, admitted that "A gentlemanly breeze from the nineteenth century drifts from the pages, bringing with it distinct déjà vu."
29. David M. Buss, "Sex Differences in Human Mate Preferences: Evolutionary Hypotheses Tested in 37 Cultures," *Behavioral and Brain Sciences* 12, no. 1 (1989): 1–14; David M. Buss, *The Evolution of Desire: Strategies of Human Mating* (New York: Basic Books, 1994); Kathryn V. Walter et al., "Sex Differences in Mate Preferences across 45 Countries: A Large-Scale Replication," *Psychological Science* 31, no. 4 (2020): 408–423.
30. The book's publisher, MIT Press, acknowledges this in the first sentences: "In this controversial book, Randy Thornhill and Craig Palmer use evolutionary biology to explain the causes of rape and to recommend new approaches to its prevention. According to Thornhill and Palmer, evolved adaptation of some sort gives rise

to rape; the main evolutionary question is whether rape is an adaptation itself or a by-product of other adaptations."

31. The "naturalistic fallacy" is the conflation of descriptive statements (what "is") with prescriptive or normative statements (what "ought" to be). Simply because something occurs in nature, does not make it good or morally right.
32. Thornhill and Palmer proposed that certain behaviors or appearances, e.g., revealing clothing, could potentially be perceived as signals of sexual availability or interest. Such "signals," they suggest, might influence the likelihood of sexual coercion or rape by men.
33. Elisabeth A. Lloyd, "Science Gone Astray: Evolution and Rape," *Michigan Law Review* 99 (2001): 1536.
34. Steven Pinker, *The Blank Slate* (New York: Penguin, 2002). Pinker lumps Richard Lewontin and Stephen Jay Gould in the "Standard Social Science Model" group along with more obvious members like anthropologists Franz Boas, Margaret Mead, and Marshall Sahlins. But his scheme seems based more on political ideology and pure opposition to sociobiology and evolutionary psychology, in the cases of Lewontin and Gould, than on academic differences.
35. Some widely used evolutionary biology textbooks barely take note of evolutionary psychology, e.g., Douglas J. Futuyma, *Evolutionary Biology*, 3rd ed. (Sunderland, MA: Sinauer Assoc., 1998); Douglas J. Futuyma and Mark Kirkpatrick, *Evolution, Fourth ed.* (Sunderland, MA: Sinauer Assoc., 2017). Evolutionary psychology has its own set of textbooks, with one of the most popular, in its sixth edition, authored by David Buss: *Evolutionary Psychology: The New Science of the Mind* (New York: Routledge, 2019), and another by Lance Workman and Will Reader, *Evolutionary Psychology: An Introduction* (Cambridge: Cambridge University Press, 2021).
36. J. A. Coyne, "Of Vice and Men," in Cheryl Brown Travis, ed., *Evolution, Gender, and Rape*, (Cambridge, MA: MIT Press, 2003).
37. For example, he points to a pivotal point in their case—that rape victims are typically of prime reproductive age—ostensibly supporting a reproductively based selection pressure. However, Coyne notes the contradictory observation that a significant number of victims are under the age of eleven, a group overrepresented in rape statistics compared with their population percentage.

Thornhill and Palmer attempt to rationalize this by referring to early onset of puberty in some under-twelve females, but this unconvincing explanation highlights the shortcomings of their hypothesis.

38. "Evolutionary psychology turns out to be pop sociobiology with a fig leaf," claim Leah Vickers and Philip Kitcher, in their own critical assessment of the evolution-of-rape hypothesis. Leah Vickers and Philip Kitcher, *Pop Sociobiology Reborn: The Evolutionary Psychology of Sex and Violence* (Cambridge, MA: MIT Press, 2003). The derisive term "pop sociobiology" comes from Kitcher's early and comprehensive critique of the field of sociobiology, particularly its application to human behavior and society, in Philip Kitcher, *Vaulting Ambition: Sociobiology and the Quest for Human Nature* (Cambridge, MA: MIT Press, 1985). Our section header borrows from Kitcher's devastating use of the quote from Shakespeare's *Macbeth*.
39. Craig T. Palmer and Randy Thornhill, "Straw Men and Fairy Tales: Evaluating Reactions to 'A Natural History of Rape,'" *Journal of Sex Research* 40 (2003): 249–55; Pinker, *The Blank Slate*, 351–379; John Alcock, *The Triumph of Sociobiology* (Oxford: Oxford University Press, 2001).
40. It is noteworthy that although most evolutionary psychology textbooks, including Buss (*Evolutionary Psychology: The New Science of the Mind*) and Workman and Reader (*Evolutionary Psychology: An Introduction*), include content on the evolution of rape in line with the Thornhill and Palmer account, others make no reference to the topic, e.g., Simon J. Hampton, *Essential Evolutionary Psychology* (Thousand Oaks, CA: Sage, 2009).
41. Owen D. Jones, "Evolutionary Analysis in Law: An Introduction and Application to Child Abuse," *North Carolina Law Review* 75 (1997): 1117.

Chapter Sixteen: The End of *Lemon* and Calls to Revisit the Synthesis

1. Elise B. Adams, "Voluntary Sterilization of Inmates for Reduced Prison Sentences," *Duke Journal of Gender Law & Policy* 26 (2018): 23; Bailey D. Barnes, "Rebalancing Judicial Immunity for Civil Rights Actions," *Tennessee Law Review* 91 (2024): 265.
2. Adams, "Voluntary Sterilization of Inmates," 28.
3. Adams, "Voluntary Sterilization of Inmates," 25.

4. *Harrington v. Purdue Pharma*, 144 S. Ct. 2071, 2078 (2024).
5. *Tangipahoa Parish Board of Education v. Freiler*, 530 U.S. 1251 (2000).
6. *Selman v. Cobb County School District*, 449 F.3d 1320 (11th Cir.).
7. Robert Shaffer, "Book Review: Lauri Lebo, *The Devil in Dover*," *Pennsylvania History: A Journal of Mid-Atlantic Studies* 76, no. 4 (2009): 213.
8. Stephen Jay Gould, "Justice Scalia's Misunderstanding," in *Constitutional Commentary* (1988): 1017; Stephen Jay Gould, "Nonoverlapping Magisteria," *Natural History* 106, no. 2 (1997), 16–22.
9. He coined the now-ubiquitous term "meme," and first used the evolutionary descriptor, "the selfish gene."
10. Richard Dawkins, *The God Delusion* (New York: Houghton Mifflin Harcourt, 2006).
11. *Torres v. Bd. of Educ. of City of Camden*, No. 101-5/10, 2010 WL 4105224 (N.J. Adm., Oct. 15, 2010).
12. *Freshwater v. Mt. Vernon City Sch. Dist.*, 137 Ohio St. 3d 469 (2013).
13. WV St. § 18-5-41a.
14. Ofer Raban, "Between Formalism and Conservatism: The Resurgent Legal Formalism of the Roberts Court," *NYU Journal of Law & Liberty* 8 (2014): 343, 344; Thomas B. Nachbar, "Twenty-First Century Formalism," *University of Miami Law Review* 75 (2020): 113, 115.
15. *Dobbs v. Jackson Women's Health Org.*, 142 S. Ct. 2228 (2022); *New York State Rifle & Pistol Ass'n, Inc. v. Bruen*, 142 S. Ct. 2111 (2022); *Sackett v. Env't Prot. Agency*, 143 S. Ct. 1322 (2023); *Students for Fair Admissions, Inc. v. President & Fellows of Harvard Coll.*, 143 S. Ct. 2141, 2154 (2023).
16. *New York State Rifle & Pistol Ass'n, Inc. v. Bruen*, 142 S. Ct. 2111 (2022).
17. *Carson v. Makin*, 596 U.S. __ (2022).
18. *Drummond v. Oklahoma Statewide Virtual Charter School Board*, __ P.3d __, 2024 WL 3155937 (OK 2024).
19. *Drummond v. Oklahoma Statewide Virtual Charter School Board*, __ P.3d __, 2024 WL 3155937 (OK 2024) (Kuehn, J., dissenting).
20. Nicholas H. Barton, "The 'New Synthesis,'" *Proceedings of the National Academy of Sciences* 119, no. 30 (2022): e2122147119. One such esoteric and unresolved detail that Barton notes: why does sexual reproduction and a "fair meiosis," i.e., equal representation of homologous chromosomes in gametes, predominate in eukaryotes (organisms whose cells have a nucleus)? Meiosis is a type of cell division that reduces the number of chromosomes in the

parent cell by half to produce four gamete cells. This reduction is crucial because it ensures that when an egg and a sperm combine during fertilization, the resulting offspring has the correct number of chromosomes. "Fair" here relates to the fact that, if there are different genetic versions (alleles) of a parental gene, each has an equal chance to be passed to offspring.

21. Richard M. Burian, "Challenges to the Evolutionary Synthesis," *Evolutionary Biology* 23 (1988): 247–269; Kevin Laland et al. "Does Evolutionary Theory Need a Rethink?," *Nature* 514, no. 7521 (2014): 161–164; Gregory A. Wray et al., "Does Evolutionary Theory Need a Rethink?—COUNTERPOINT No, All Is Well," *Nature* 514, no. 7521 (2014).
22. Successful academic researchers essentially brand themselves—they depend on name recognition. Thus Lala's decision to change his name, well into his career, was somewhat unusual. Lala explains that his parents had Anglicized the original family name when he was four years old in an attempt to reduce the racism their children might experience. But he decided, at age 54, to celebrate his ancestry and return to the original family name. Lala clearly is not someone to be intimidated by a challenge to conventional wisdom. He is a prolific researcher, with some 230 scientific articles and multiple books, whose research interest targets evolution and animal behavior in areas such as "niche construction."
23. Kevin Laland, Tobias Uller, Marc Feldman, Kim Sterelny, Gerd B. Müller, Armin Moczek, Eva Jablonka, et al., "Does Evolutionary Theory Need a Rethink?," *Nature* 514, no. 7521 (2014): 161–164.
24. Stephen Buranyi, "Do We Need a New Theory of Evolution?," *The Guardian*, June 28, 2022, accessed January 8, 2024, https://www.theguardian.com/science/2022/jun/28/do-we-need-a-new-theory-of-evolution.
25. Including contentious issues we considered such as punctuated equilibrium, certain claims in sociobiology and evolutionary psychology.
26. Laland, "Does Evolutionary Theory Need a Rethink?," 161–164; Kevin Laland, "The Extended Evolutionary Synthesis: Its Structure, Assumptions and Predictions," *Proceedings of the Royal Society B: Biological Sciences* 282, no. 1813 (2015): 20151019. We focus on these publications because they have received specific attention in the popular press, but there are other significant players in this

arena: Massimo Pigliucci, "Do We Need an Extended Evolutionary Synthesis?," *Evolution* 61, no. 12 (2007): 2743–2749; Massimo Pigliucci, "An Extended Synthesis for Evolutionary Biology," *Annals of the New York Academy of Sciences* 1168, no. 1 (2009): 218–228; Massimo Pigliucci and Gerd B. Muller, *Evolution—The Extended Synthesis* (Cambridge, MA: MIT Press, 2010).

27. Which, to modern evolutionary biologists, includes sexual selection as a subset of natural selection.
28. Laland, "The Extended Evolutionary Synthesis," 5.
29. Synthesists (Mayr, for example) often used the term "epigenetic" in reference to developmental processes, not in the modern sense of molecular DNA expression.
30. Wray, "Does Evolutionary Theory Need a Rethink?—COUNTERPOINT."
31. Douglas J. Futuyma wrote a more extensive rebuttal to the EES proposal in 2017. Futuyma is professor of ecology and evolution at the State University of New York at Stony Brook and winner of the Sewall Wright Award from the American Society of Naturalists, as well as the authoritative author of the most widely used undergraduate textbooks on evolution. Futuyma echoed the Traditionalist response and concluded that the suggested expansions to the Synthesis are unnecessary. He contends that all the research cited by EES proponents is satisfactorily accommodated within the existing framework, reflecting its adaptability and robustness. Douglas J. Futuyma, "Evolutionary Biology Today and the Call for an Extended Synthesis," *Interface Focus* 7 (2017): 20160145
32. *Peloza v. Capistrano Unified Sch. Dist.*, 782 F. Supp. 1412, 1415 (C.D. Cal. 1992).
33. *Colbert v. Infinity Broad. Corp.*, 423 F. Supp. 2d 575, 585–86 (N.D. Tex. 2005).
34. Joelle Anne Moreno, "Extralegal Supreme Court Policy-Making," *William & Mary Bill of Rights Journal* 24, no. 2 (2015): 451–520.
35. *Finkbeiner v. Geisinger Clinic*, 623 F. Supp. 3d 458, 463 (M.D. Pa. 2022).
36. *Berutti v. Wolfson*, 2023 WL 1071624 (D.N.J. Jan. 27, 2003).

Conclusion

1. Daniel C. Dennett, "Real Patterns," *Journal of Philosophy* 88, no. 1 (1991): 27–51; Daniel C. Dennett, *Darwin's Dangerous Idea: Evolution and the Meanings of Life* (New York: Simon & Schuster, 1995);

Daniel C. Dennett, *From Bacteria to Bach and Back: The Evolution of Minds* (New York: W.W. Norton & Company, 2017).

2. V. Masson-Delmotte, P. Zhai, A. Pirani, S. L. Connors, C. Péan, S. Berger, N. Caud, Y. Chen, L. Goldfarb, M. I. Gomis, M. Huang, K. Leitzell, E. Lonnoy, J. B. R. Matthews, T. K. Maycock, T. Waterfield, O. Yelekçi, R. Yu, and B. Zhou, eds., *Climate Change 2021: The Physical Science Basis. Contribution of Working Group I to the Sixth Assessment Report of the Intergovernmental Panel on Climate Change*, IPCC, 2021: Summary for Policymakers (Cambridge: Cambridge University Press, 2021), 3–32, doi:10.1017/9781009157896.001.
3. Mark Lynas et al., "Greater than 99% Consensus on Human Caused Climate Change in the Peer-Reviewed Scientific Literature," *Environmental Research Letters* 16 (2021): 114005.
4. In the legal field, for example, premier scholars have accepted the evidence-based conclusions of climate scientists, e.g., Michael J. Klarman, "Foreword: The Degradation of American Democracy-and the Court," *Harvard Law Review* 134 (2020): 1, 175 (referencing the "copious evidence decimating" theories of climate-change denialism). Nevertheless, some Court decisions still refer to climate change as a controversial question. *Janus v. AFSCME*, 585 U.S. 878, 913 (2018); *National Rev., Inc. v. Mann*, 140 S. Ct. 344, 346 (2019) (Alito, J., dissenting from denial of cert.).
5. See *Matter of Hawai'i Elec. Light Co., Inc.*, __ P.3d __, No. SCOT-22-418, 2023 WL 2471890, at *6 (Haw. Mar. 13, 2023) (considering climate implications of state constitutional "right to a life-sustaining climate system"); *Vereniging Milieudefensie v. Royal Dutch Shell PLC*, Hague District Court, Judgment of May 26, 2021 (holding that Royal Dutch Shell owed a standard of care to the community, requiring it to limit greenhouse gas emissions).
6. Scott Novak, "The Role of Courts in Remedying Climate Chaos: Transcending Judicial Nihilism and Taking Survival Seriously," *Georgetown Environmental Law Review* 32 (2020): 743, 747; Katrina Fischer Kuh, "The Legitimacy of Judicial Climate Engagement," *Ecological Law Quarterly* 46 (2019): 731.
7. Nathan Richardson, "The Rise and Fall of Clean Air Act Climate Policy," *Michigan Journal of Environmental & Administrative Law* 10 (2020): 69, 71–73; Karen C. Sokol, "Seeking (Some) Climate Justice in State Tort Law," *Washington Law Review* 95 (2020): 1383, 1417–1434.

8. Naomi Oreskes and Erik M. Conway, *Merchants of Doubt: How a Handful of Scientists Obscured the Truth on Issues from Tobacco Smoke to Global Warming* (London: Bloomsbury, 2010).
9. In the spring of 2017, HG received an unsolicited package from the Heartland Institute. As detailed in *Merchants of Doubt*, Heartland worked with the tobacco industry in the 1990s to question the evidence about cancer risks from secondhand smoke and, in 2007, began challenging the data on climate change, and whether or not there was indeed scientific consensus about it. The package contained a memo sent to "Professors of Physical Science," describing the contents and the mission of the Institute. In addition to the memo, there was a slim but slickly produced hundred and ten-page book, authored by a trio of climate change deniers, including S. Fred Singer, one of the featured "merchants of doubt" in Oreskes and Conway's book. In chapters with titles such as "No Consensus," "Why Scientists Disagree," "Flawed Projections," "False Postulates," and "Unreliable Circumstantial Evidence," the book attempts to persuade the reader that climate change is not real. HG was not convinced. Craig D. Idso, Robert M. Carter, and S. Fred Singer, *Why Scientists Disagree about Global Warming*, 2nd edition (Arlington Heights, IL: Heartland Institute, 2016).
10. Naomi Oreskes and Erik M. Conway, "Defeating the Merchants of Doubt," *Nature* 465, no. 7299 (2010): 686–687.
11. Megan Brenan, "40% of Americans Believe in Creationism," *Gallup*, July 26, 2019, https://news.gallup.com/poll/261680/americans-believe-creationism.aspx
12. Oreskes and Conway cite a December 2009 Angus Reid poll revealing that only 44% of Americans agreed that "global warming is a fact and is mostly caused by emissions from vehicles and industrial facilities" and that "[t]here has been essentially no change in public acceptance of the scientific conclusions since the 1980s." Oreskes and Conway, "Defeating the Merchants of Doubt," 686–687.
13. In 2012, *The Double Helix* was included as one of the eighty-eight "Books That Shaped America" by the Library of Congress. James D. Watson, *The Double Helix: A Personal Account of the Discovery of the Structure of DNA* (New York: Atheneum Press, 1968).
14. Walter Isaacson, *The Code Breaker: Jennifer Doudna, Gene Editing, and the Future of the Human Race* (New York: Simon & Schuster, 2021).

Issacson's account of the competition involved in the discovery of the gene editing tool, CRISPR, reveals this component of science is still very much in place.

15. Norman A. Johnson, *Darwin's Reach: 21st Century Applications of Evolutionary Biology* (Boca Raton, FL: CRC Press, 2021). Johnson points to how evolutionary biology is profoundly relevant to diverse problems in healthcare and medicine, environmental science (including climate change), agriculture, and sustainable energy.
16. Jeffrey M. Jones, "Supreme Court Approval Holds at Record Lows," *Gallup*, August 2, 2023.
17. Stephen Feldman, "Saving the Supreme Court? Constitutional Rights and the Inevitability of Politics (with a Discussion of Antisemitism)," *Houston Law Review* 61 (2024): 953, 956.
18. James J. Sample, "The Supreme Court and the Limits of Human Impartiality," *Hofstra Law Review* 52 (2024): 579; Laurie L. Levenson, "The Word Is 'Humility': Why the Supreme Court Needed to Adopt a Code of Judicial Ethics," *Pepperdine Law Review* 51 (2024): 515.
19. Lisa Avalos, "The Innocence Standard: Supreme Court Nominees and Sexual Misconduct," *Connecticut Law Review* 56 (2024): 345.
20. Elizabeth D. Walker, "Public Confidence in the Judiciary," *West Virginia Law Review* (Summer 2023): 18, https://www.courtswv.gov/sites/default/pubfilesmnt/2023-06/2023_04to06_ChiefColumn_1.pdf.
21. Jeff Tollefson, "The Supreme Court's War on Science," *Nature* 609 (2022): 460–462.
22. Rhona Mijumbi, "Science versus Government: Can the Power Struggle Ever End?," *Nature* 626 (2024): 256–257.
23. Geoff Mulgan, *When Science Meets Power* (Cambridge: Polity Press, 2023).

Epilogue

1. *Roake v. Brumley*, No. CV 24-517, 2024 WL 4746342, at *1 (M.D. La. Nov. 12, 2024).
2. *Alliance for Hippocratic Medicine v. FDA*, 668 F. Supp. 3d 507, 539 (N.D. Tex.), vacated and remanded, 117 F.4th 336 (5th Cir. 2024).
3. *Alliance for Hippocratic Medicine v. FDA*, 78 F.4th 210, 259 (5th Cir.), rev'd and remanded, 602 U.S. 367 (2024).

INDEX